T0235776

Cyber-Physical Vehicle Systems

Methodology and Applications

Synthesis Lectures on Advances in Automotive Technologies

Editor
Amir Khajepour, *University of Waterloo*

The automotive industry has entered a transformational period that will see an unprecedented evolution in the technological capabilities of vehicles. Significant advances in new manufacturing techniques, low-cost sensors, high processing power, and ubiquitous real-time access to information mean that vehicles are rapidly changing and growing in complexity. These new technologies—including the inevitable evolution toward autonomous vehicles—will ultimately deliver substantial benefits to drivers, passengers, and the environment. Synthesis Lectures on Advances in Automotive Technology Series is intended to introduce such new transformational technologies in the automotive industry to its readers.

Real-Time Road Profile Identification and Monitoring: Theory and Application
Yechen Qin, Hong Wang, Yanjun Huang, and Xiaolin Tang
2018

Noise and Torsional Vibration Analysis of Hybrid Vehicles
Xiaolin Tang, Yanjun Huang, Hong Wang, and Yechen Qin
2018

Smart Charging and Anti-Idling Systems
Yanjun Huang, Soheil Mohagheghi Fard, Milad Khazraee, Hong Wang, and Amir Khajepour
2018

Design and Avanced Robust Chassis Dynamics Control for X-by-Wire Unmanned Ground Vehicle
Jun Ni, Jibin Hu, and Changle Xiang
2018

Electrification of Heavy-Duty Construction Vehicles
Hong Wang, Yanjun Huang, Amir Khajepour, and Chuan Hu
2017

Vehicle Suspension System Technology and Design
Avesta Goodarzi and Amir Khajepour
2017

Cyber-Physical Vehicle Systems: Methodology and Applications

Chen Lv, Yang Xing, Junzhi Zhang, and Dongpu Cao

ISBN: 978-3-031-00376-9   paperback
ISBN: 978-3-031-01504-5   ebook
ISBN: 978-3-031-00009-6   hardcover

DOI 10.1007/978-3-031-01504-5

A Publication in the Springer series
SYNTHESIS LECTURES ON ADVANCES IN AUTOMOTIVE TECHNOLOGIES

Lecture #10
Series Editor: Amir Khajepour, *University of Waterloo*
Series ISSN
Print 2576-8107   Electronic 2576-8131

Cyber-Physical Vehicle Systems

Methodology and Applications

Chen Lv
Nanyang Technological University, Singapore

Yang Xing
Nanyang Technological University, Singapore

Junzhi Zhang
Tsinghua University, P.R. China

Dongpu Cao
University of Waterloo

SYNTHESIS LECTURES ON ADVANCES IN AUTOMOTIVE TECHNOLOGIES #10

ABSTRACT

This book studies the design optimization, state estimation, and advanced control methods for cyber-physical vehicle systems (CPVS) and their applications in real-world automotive systems. First, in Chapter 1, key challenges and state-of-the-art of vehicle design and control in the context of cyber-physical systems are introduced. In Chapter 2, a cyber-physical system (CPS) based framework is proposed for high-level co-design optimization of the plant and controller parameters for CPVS, in view of vehicle's dynamic performance, drivability, and energy along with different driving styles. System description, requirements, constraints, optimization objectives, and methodology are investigated. In Chapter 3, an Artificial-Neural-Network-based estimation method is studied for accurate state estimation of CPVS. In Chapter 4, a high-precision controller is designed for a safety-critical CPVS. The detailed control synthesis and experimental validation are presented. The application results presented throughout the book validate the feasibility and effectiveness of the proposed theoretical methods of design, estimation, control, and optimization for cyber-physical vehicle systems.

KEYWORDS

cyber-physical vehicle systems, co-design optimization, dynamic modeling, design space exploration, parameter optimization, state estimation, neural networks, controller synthesis, simulation validation, experimental testing

Contents

Preface

This book studies the design optimization, state estimation, and advanced control methods for Cyber-Physical Vehicle Systems (CPVS) and their applications in real-world automotive systems.

In Chapter 1, key challenges and state-of-the-art of vehicle design and control in the context of cyber-physical systems are introduced. In Chapter 2, a Cyber-Physical System (CPS)-based framework for co-design optimization of an automated electric vehicle with different driving styles was proposed. The multi-objective optimization problem was formulated. The driving style recognition algorithm was developed using unsupervised machine learning and validated via vehicle testing. The system modelling and experimental verification were carried out. Vehicle control algorithms were synthesized for three typical driving styles with different protocol selections. The performance exploration methodology and algorithms were proposed. Test results show that the overall performances of the vehicle were significantly improved by the proposed co-design optimization approach. Future work will be focused on real vehicle application of the proposed methods and CPS design methodology improvement.

In Chapter 3, a novel probabilistic estimation method of brake pressure is developed for a safety critical CPVS based on multilayer Artificial Neural Network (ANN) with Levenberg-Marquardt Backpropagation Traning (LMBP) training algorithm. The high-level architecture of the proposed multilayer ANN for brake pressure estimation is illustrated at first. Then, an efficient algorithm of LMBP method is developed for model training. The real vehicle testing is carried out on a chassis dynamometer under New European Drive Cycle (NEDC) driving cycles. The experimental results show that the developed model can accurately estimate the brake pressure, and its performance is advantageous over other learning-based methods with respect to estimation accuracy, demonstrating the feasibility and effectiveness of the proposed algorithm.

In Chapter 4, a typical safety-critical CPVS, i.e., the Brake-By Wire (BBW) system, was introduced. Compared to the existing BBW system, the newly developed system enjoys the advantage of a simple structure and low cost because only conventional valves and sensors are added to the usual hydraulic layouts. Two pressure modulation methods, namely, the Hydraulic Pump-Based Pressure Modulation (HPBPM) and Closed-Loop Pressure-Difference-Limiting (CLPDL) modulation, were proposed to improve the modulation precision of hydraulic brake pressure and reduce valve's operation noise as well. Experiments were conducted in a hardware-in-the-loop test rig to demonstrate the performance of the proposed control methods.

Chen Lv, Yang Xing, Junzhi Zhang, and Dongpu Cao
December 2019

CHAPTER 1

Introductions

Intelligent vehicles have been gaining increasing attention from both academia and industrial sectors [1]. The field of intelligent vehicles exhibits a multidisciplinary nature, involving transportation, automotive engineering, information, energy, and security [2–5]. Intelligent vehicles have increased their capabilities in highly and even fully automated driving. However, unresolved problems do exist due to strong uncertainties and complex human-vehicle interactions.

Highly automated vehicles are likely to be on public roads within a few years. Before transitioning to fully autonomous driving, driver behavior should be better understood and integrated to enhance vehicle performance and traffic efficiency [6–9]. To address these challenges, researchers have explored advanced driver assistance systems (ADAS), and human-machine interface (HMI) from a variety of points of view [10, 11]. However, since the dynamic relationships between driver and vehicle are highly complex, satisfactory driver-vehicle interactions should go beyond the present ADAS and HMI systems. Human-vehicle interactions have already been considered in a high-level closed loop, where driving style, driving feel, and vehicle performance are considered [12]. Driving style plays a very important role in vehicle energy efficiency and ride comfort, thus significantly impacting controller synthesis [12–14]. For instance, control objectives and control protocols should be adaptively adjusted according to different driving styles. Based on the findings reported in [13], a better understanding of driving styles could help improve ADAS performance and further reduce vehicle's fuel consumption through driver feedback. In [14], an enhanced intelligent driver model was developed, and then it was used to investigate the impact of different driving strategies on traffic capacity. In [15], an adaptive cruise control strategy considering the characteristics of different driving styles was developed, and the proposed strategy could automatically adapt to different traffic situations. Nevertheless, advanced control and optimization of vehicle systems with characterized driving styles are still open challenges and worthwhile exploring.

In the meantime, the ever-growing attention to the environment and energy conservation requires automobiles to be cleaner and more efficient [16–18]. In this study, an electric vehicle (EV) is chosen as the platform to conduct our research in cyber-physical vehicle systems. Based on existing studies, small changes in driving style can cause unnecessary energy waste and suboptimal performance of an EV [19, 20]. Moreover, regenerative braking capability of EVs can be enhanced by prior knowledge of driving style. Hence, an optimal energy management strategy can be obtained with knowledge about the entire driving cycle, environment, and driver behaviors. Therefore, the information of operating scenarios, driver behaviors, and driver-vehicle

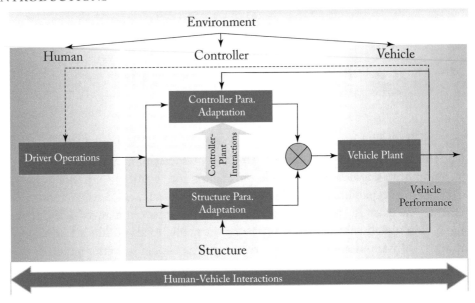

Figure 1.1: Schematic diagram of the CPVS.

interactions are crucial and should be integrated to enhance the energy efficiency of automated electric vehicles.

A Cyber-Physical System (CPS) is a distributed, networked system that fuses computational processes (cyber world) with the physical world. An intelligent electric vehicle is a typical example of Cyber-Physical Vehicle System (CPVS). In details, an automated electric vehicle involves the following subsystems: the controller, representing the "Cyber" world, the physical vehicle plant, the driver, the "Human," and the environment. These different parts, which are highly coupled, decide the vehicle's behavior and final performance, as Fig. 1.1 shows. The main drawback of the conventional implementations in vehicle design and control is the lack of global optimality in the selection of architecture, parameters, and variables [25]. For instance, by using the conventional design method, which deals with different subsystems independently, even if the controller is very well designed, the improvement of vehicle performance could be limited, since the physical architecture and parameters are not optimized in sync with the controller, and the system potential is not fully explored. In this context, the emerging co-design method provides the capability to extend system design space and further enhance the performance of CPS [24–28]. In [24], a platform-based design method utilizing contracts to do the high-level abstraction of the components in a CPS was proposed, and it is able to offer support to the overall design process. In [26], co-design optimization of a cyber physical vehicle system, which considers task time, actuator characteristics, energy consumption, and processor workload, was investigated. In [27], a CPS-based control framework was developed for vehicle systems to minimize the car-following fuel consumption and ensure inter-vehicle safety. Besides the cyber and the physical worlds, we also need to take "Human" of an automated vehicle into consideration. Thus, the interactive impacts between the vehicle plant, control variables, multi-performance, and driver styles should be well understood [29–31].

To further advance the existing CPS methods as well as their applications in vehicle engineering, the following topics will be explored for CPVS in this book: (1) a novel co-design optimization methodology for CPVS; (2) dynamic estimation of hybrid states for online monitoring of CPVS; and (3) advanced control synthesis for CPVS for improving multiple performance.

CHAPTER 2

Co-Design Optimization for Cyber-Physical Vehicle System

2.1 PROBLEM FORMULATION

In this section, the co-design of a typical CPVS, i.e., an automated electric vehicle, with different driving styles is formulated as a multi-objective optimization problem. The goal is to find optimal assignments for design variables to maximize performances while satisfying a number of constraints. To ensure the problem to be solved within a reasonable complexity, the following assumptions are made: (1) the vehicle operates in normal conditions, and vehicle stability could be guaranteed by stability control functions; (2) only longitudinal motion control is considered in this study; and (3) the sizing of the electric powertrain is fixed, i.e., the parameters of the battery and the electric motor are constant to bound the exploration space.

2.1.1 HIERARCHICAL OPTIMIZATION METHODOLOGY

The optimization problem is formulated as a constrained multi-objective one where both vehicle and controller parameters need to be chosen. In this book, the Platform-Based Design (PBD) is adopted as the co-design methodology [21].

As Fig. 2.1 shows, PBD is a meet-in-the-middle approach that favors re-usability. At the top layer, there are high-level requirements and constraints. The bottom layer is defined by a design *platform*, i.e., a library of components characterized by their behaviors and performance. In this study, the bottom layer contains the models of the vehicle, electric powertrain, brakes, and driver-style-based controller. The models are parametrized to capture families of the system, components and controllers. The design problem is to select a set of components and their parameters so that the constraints are satisfied with the objective functions optimized. The selection process is called mapping, indicated as the middle-layer meeting point in the diagram, since the obligations captured in the requirements and constraints are discharged by particular components or combinations thereof. Co-design of the physical parameters, controller protocols, and variables for the intelligent electric vehicle is then made possible.

2.1.2 SYSTEM DESCRIPTION

(1) *Physical plant:* For the structure of the studied automated electric vehicle, a central electric motor is installed at the front axle of the vehicle. During acceleration, the motor, which is

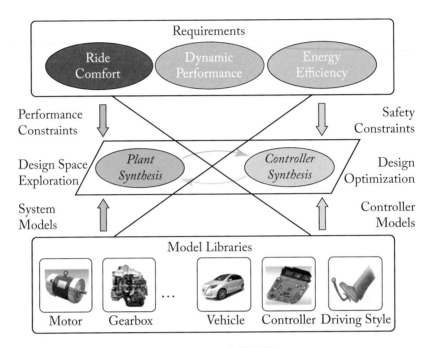

Figure 2.1: Platform-based design optimization of CPVS.

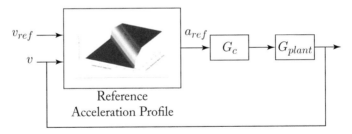

Figure 2.2: Longitudinal motion control architecture of the intelligent vehicle.

powered by the battery, provides propulsion through the transmission system to the wheels. During deceleration, the regenerative braking torque generated by the motor is blended with the friction braking modulated by the hydraulic modulator.

(2) *Control architecture:* The high-level strategy for the longitudinal motion control of the automated EV is designed to track a reference acceleration, generated via the pre-defined acceleration profile, as shown in Fig. 2.2. The reference acceleration profile is a 3D look-up table defined by the reference vehicle speed v_{ref}, the ego-vehicle speed v, and the reference acceleration a_{ref}.

2.1.3 DRIVING EVENT

A *driving event* is a driving maneuver, such as acceleration, deceleration, turning, and lane change, which can be used to identify driving styles [28]. As mentioned previously, this study mainly focuses on longitudinal motion control, hence the adopted driving events are defined as [29] follows.

(1) *Event 1: 0–50 km/h acceleration.* In this event, the car is accelerated from 0–50 km/h. The vehicle acceleration, jerk, and the time taken in this process are typical performance indices. This event is used to optimize and evaluate the dynamic performance and ride comfort under different driving styles.

(2) *Event 2: 50–0 km/h deceleration.* In this event, the car is decelerated from 50 km/h to 0. The deceleration and the time taken in this process are typical performance indices. The energy recovered during the braking process can be used to evaluate energy efficiency. This event is used to optimize and check vehicle's dynamic performance and energy efficiency under different driving styles.

(3) *Event 3: driving cycle.* Although the energy consumption of the vehicle can be evaluated in the above two events, the time duration of an acceleration or deceleration procedure is relatively short, making it difficult to evaluate energy consumption at the vehicle level. Thus, the ECE driving cycle is adopted for measuring energy efficiency under different driving styles. The ECE driving cycle, which is a series of data points representing the vehicle speed vs. time, exhibits the typical driving conditions of a car in urban areas [17]. It is usually adopted to carry out road testing for studying the fuel economy of a passenger car.

2.1.4 DRIVING STYLE RECOGNITION

To identify driving style for control synthesis and system optimization, a driving style recognition (DSR) algorithm is developed using unsupervised machine learning with partially labeled data. The data set is collected in the road tests with a Sedan-Type vehicle, and it is comprised of 9 real life cycles covering over 500 km. The data can be overall classified into three groups according to the driver feedback as aggressive, conservative, and moderate. These three driving styles are firstly defined as [29–34] as follows.

(1) *Aggressive:* Aggressive drivers exhibit frequent changes in throttle and brake pedal positions [32]. They drive with sharp and abrupt accelerations and decelerations, aiming at vehicle dynamic performance. This kind of behavior would result in higher fuel consumption and increased likelihood of accidents [29].

(2) *Conservative:* Conservative drivers often exhibit mild operational behaviors with small amplitudes and low-frequency actions on a steering wheel, accelerator, and brake pedal [33]. They value energy efficiency and ride comfort, and avoid abrupt variations of vehicle state.

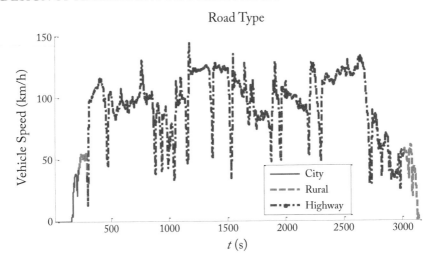

Figure 2.3: The real life route used for DSR experimental validation.

(3) *Moderate:* Moderate drivers are positioned between the above two. They would like to balance multiple performances, such as vehicle dynamic performance, ride comfort, and energy efficiency [29].

The unlabeled data set is pre-processed for driving events detection and statistics extraction. A total amount of six signals is used: throttle pedal position, brake light switch, longitudinal and lateral accelerations, steering wheel angle and vehicle speed. Five statistics are extracted per event: maximum, minimum, mean, standard deviation, and root mean square. The reduced set of signals is clustered using Gaussian Mixture Models (GMM), which generates the DSR classification algorithm to be implemented onboard. The performance of the DSR algorithm is validated against the subjective labels and further tested with a new set of data from a new real-life route with changeable road type, as shown in Fig. 2.3. This new data set is collected by a Sport Utility Vehicle (SUV)-type vehicle with a different driver.

Table 2.1 shows the results of the SUV driving data using the developed DSR algorithm. So as to quantitatively evaluate the performance of the algorithm, the driving cycles are classified per events using the aggressiveness index. The aggressiveness index is transformed from the classification into an equivalent index, assigning an increasing value from 0–1 to the different events based on the level of aggressiveness [34]. To provide further information about the robustness of each classification, the number of events identified is included in brackets and italics. According to the results, the conservative cycle is classified as the least aggressive one, particularly by acceleration and brake events analysis. The moderate cycle is situated between the aggressive and conservative ones. While the aggressive cycle is identified as the sportiest one, but it has a similar braking level with the moderate one, agreeing with driver's feedback.

Table 2.1: Driving style recognition results in suv cycles

	Aggressive Cycle	Moderate Cycle	Conservative Cycle
Acceleration	0.55 (*149*)	0.43 (*113*)	0.34 (*106*)
Brake	0.58 (*33*)	0.56 (*25*)	0.36 (*22*)
Cruise	0.83 (*149*)	0.69 (*126*)	0.70 (*124*)
Turn	0.41 (*6*)	0.29 (*7*)	0.29 (*7*)

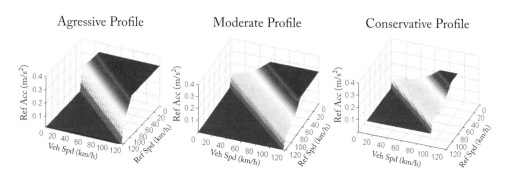

Figure 2.4: Pre-defined 3D reference acceleration profiles.

Finally, the consistency and robustness of the algorithm are verified using the test data set. The test shows consistency in the identification and aligns with drivers' perception. The above testing results validate the suitability of this approach for DSR, its onboard implement capability and robustness to vehicle and driver characteristics. More detailed algorithms with experimental results can be found in [34].

Based on the above recognition and classification algorithms, the features of aggressive, conservative, and moderate driving styles can be extracted, and online recognition of a driver's driving style can be realized using the well-trained model as well. Meanwhile, according to the above features obtained, the three dimensional human-like acceleration profiles are developed for each driving style, as illustrated in Fig. 2.4.

2.1.5 REQUIREMENTS FOR THE DESIGN AND OPTIMIZATION OF CPVS

The requirements for vehicle design and control involve dynamical performance, energy efficiency, and ride comfort. Driving style consideration implies the introduction of multiple trade-offs between performances that are set as the objective functions in our optimization problem, under different driving styles, operating conditions, and driving tasks.

(1) *Dynamic performance:* Dynamic performance is considered as the fundamental and the most important indicator of a car [29]. Maximum speed and acceleration time are proxies for dynamic performance. In this study, we select the 0–50 km/h acceleration time t_{acc} and the 50–0 km/h deceleration time t_{brk} as two indicators for the dynamic performance to capture driver's behavior and select suitable value for the gear ratio i_g.

(2) *Ride comfort:* The comfort level of a vehicle, also known as drivability, can be assessed by vehicle's jerk j, which is the second derivative of the vehicle's longitudinal velocity v [17]:

$$j = \ddot{v}. \tag{2.1}$$

During acceleration, torsional oscillations may occur in the drivetrain due to fast torque transitions, resulting in unexpected jerks at vehicle level and deteriorated drivability. To cope with this problem, an active damping controller is usually required [36]. Although aggressive drivers may enjoy fierce acceleration and jerk, for those who prefer conservative or moderate driving style, ride comfort is a very important performance. In this study, jerk is used to capture the comfort level of the vehicle.

(3) *Energy efficiency:* The energy efficiency of a vehicle can be represented by the energy consumed during a certain trip. Typically, energy consumption can be reduced by optimizing the powertrain energy management [29]. For electrified vehicles, it can be further enhanced through regenerative braking. Thus, in this study, the regenerated braking energy defined in Equation (2.2) is set as one of the optimization goals in the trade-off problem [18]:

$$E_{reg} = \eta_{gen} \cdot \int T_{m,reg} \omega_m dt, \tag{2.2}$$

where E_{reg} is the regenerated brakin energy, $T_{m,reg}$ and ω_m are the regenerative braking torque and the angular speed of the electric motor, respectively, and η_{gen} is the generation efficiency of the motor.

2.1.6   CONSTRAINTS FOR VEHICLE DESIGN AND OPTIMIZATION

Constraints in the optimization problem involve indicators that are set to stay within specific bounds to limit the search space.

(1) *Maximum vehicle speed:* The constraint on vehicle speed is posed as:

$$v_{\max} = r \pi n_{\max}/(30 i_g) \geq (100/3.6) \text{ m/s}, \tag{2.3}$$

where $v_{\max}$ is the maximum speed of the vehicle, $n_{\max}$ is the highest rotational speed of the electric motor, r is the nominal radius of tire, and i_g is the gear ratio.

(2) *Minimum gradeability: Gradeability* is defined as the highest grade that a vehicle can achieve with a maintained speed. Once the motor parameters are given, this performance is

determined by the gear ratio, as Equation (2.4) shows [35]:

$$\eta_t i_g T_{m,max} = mgr\,(f\cos\alpha_{max} + \sin\alpha_{max}) \tag{2.4}$$

$$i_{max} = \tan\alpha_{max} \geq 30\%, \tag{2.5}$$

where $T_{m,max}$ is motor's peak torque, m is the total mass of the vehicle, η_t is the efficiency of the transmission system, f is the friction drag coefficient, and α is the grade angle.

(3) *Minimum brake intensity:* In order to guarantee stability during braking, a vehicle needs to have enough braking force, represented by the brake intensity z, as required by regulation ECE-R13 [36]:

$$z = \dot{v}/g \geq 0.1 + 0.85(\varphi - 0.2), \tag{2.6}$$

where φ is the adhesion coefficient of the road.

(4) *Powertrain limits:* According to the assumption described above, the characteristics of the power source are given, then the limitation on motor torque can be described by:

$$T_m \omega_m \leq P_{m,lim}, \tag{2.7}$$

where T_m is output torque of the electric motor, and $P_{m,lim}$ is the peak power of the electric motor.

2.2   SYSTEM MODELING AND VALIDATION

2.2.1   ELECTRIC POWERTRAIN SYSTEM

The electric powertrain is comprised of an electric motor, a gearbox, a final drive, a differential, and half shafts. The motor torque is modeled as a first-order reaction, as shown in Equation (2.8). The models for the drivetrain dynamics and half-shaft torque can be given by Equations (2.9) and (2.10) [25]:

$$T_{m,ref} = T_m + \tau_m \dot{T}_m \tag{2.8}$$

$$J_m \ddot{\theta}_m = T_m - 2T_{hs}/i_g \tag{2.9}$$

$$T_{hs} = k_{hs}\left(\theta_m/i_g - \theta_w\right) + c_{hs}\left(\dot{\theta}_m/i_g - \dot{\theta}_w\right), \tag{2.10}$$

where τ_m is the small time constant, $T_{m,ref}$ is the reference torque of the electric motor, T_{hs} is the half-shaft torque, J_m is the motor inertia, and θ_m and θ_w are the angular positions of electric motor and load, respectively. k_{hs} and c_{hs} are the stiffness and damping coefficients of the half shaft, respectively.

In this study, the battery is built as an open-circuit voltage-resistance model. Look-up tables are compiled on the basis of the state of charge (SOC) and temperature data of the battery, modeling its charging-discharging internal resistance. The detailed model with parameters can be found in [17].

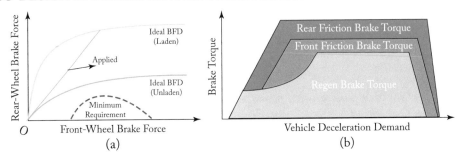

Figure 2.5: Brake force distribution strategy.

2.2.2 BLENDED BRAKE SYSTEM

The brake force distribution (BFD) should adhere to the ideal curve. To simplify the implementation and to avoid real-time modulation of brake pressure, the BFD is usually set as a fixed value, which is determined by the parameters of the installed brake devices, as shown in Fig. 2.5a. The front and rear braking demands can be calculated as follows [17]:

$$T_b = 2T_{b,fw} + 2T_{b,rw} \tag{2.11}$$

$$T_{b,fw} = \beta T_{b,dmd}/2 \tag{2.12}$$

$$T_{b,rw} = (1 - \beta)T_{b,dmd}/2, \tag{2.13}$$

where T_b is the actual braking torque provided by the blended brakes, $T_{b,dmd}$ is the demanded braking torque of the vehicle, and $T_{b,fw}$ and $T_{b,rw}$ are the requested braking torque of one front wheel and one rear wheel, respectively. β is the BFD ratio.

As shown in Fig. 2.5b, during deceleration, the overall demanded braking torque of the vehicle is supplied by the regenerative and the friction blending braking. The overall braking torque is controlled to be consistent with driver's deceleration intention. The reference values for the regenerative and frictional braking on front axle can be given by:

$$T_{m,reg} = \min\left(2T_{b,fw}/i_g, T_{m,reg,lim}\right) \tag{2.14}$$

$$T_{b,fw,fric} = T_{b,fw} - 2T_{m,reg}/i_g, \tag{2.15}$$

where $T_{m,reg}$ and $T_{m,reg,lim}$ are reference torque and torque limit of the regenerative braking of the electric motor, respectively. $T_{b,fw,fric}$ is the frictional braking torque of the front wheel.

2.2.3 DYNAMIC MODEL OF THE VEHICLE AND TYRE

A model of vehicle dynamics with seven degrees of freedom has been built. The tyre model, which is of great importance for research on acceleration and deceleration, should be able to simulate the real tyre in both adhesion and sliding. In this article, the well-known Pacejka magic formula tyre model is adopted [37]. The detailed models were described in [17].

Table 2.2: Key parameters of the electric vehicle

Parameter	Value	Unit
Vehicle mass	1360	kg
Wheel base	2.50	m
Frontal area	2.40	m^2
Gear ratio	7.881	—
Nominal radius of tire	0.295	m
Coefficient of air resistance	0.32	—
Motor peak power	45	kW
Motor maximum torque	145	Nm
Motor maximum speed	9000	rpm
Battery voltage	336	V
Battery capacity	66	Ah

2.2.4   EXPERIMENTAL VALIDATION

The models of the electric vehicle with its subsystems were implemented in MAT-LAB/Simulink. Experimental data measured from vehicle test were used for model calibration. Key parameters of the systems are listed in Table 2.2. The feasibility and effectiveness of the models have been previously validated via hardware-in-the-loop experiments and vehicle road testing [17, 25].

2.3   CONTROLLER DESIGN FOR DIFFERENT DRIVING STYLES

2.3.1   HIGH-LEVEL CONTROLLER ARCHITECTURE

The high-level supervisory controller adopts a scheduling protocol, asking the architecture and control objectives of the low-level controller, as well as the parameters of the physical plant, to dynamically adapt to different driving styles, as shown in Fig. 2.6. In this study, the driving style of the automated vehicle can be either obtained in the manual mode through the DSR algorithm developed in the previous section, or actively selected by human operator during autonomous mode. To avoid unexpected discontinuities in controller output resulted by frequent and fast transitions between different driving styles, a simple and reliable approach for the application is to allow the driving style to be actively or passively switched only when the vehicle is stopped, i.e., the vehicle speed $v = 0$.

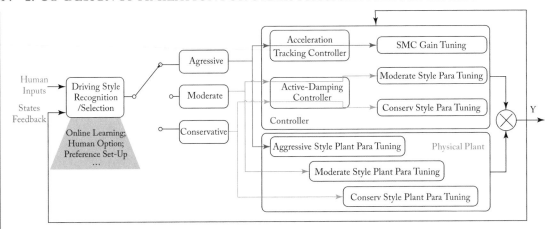

Figure 2.6: Scheduling-protocol based hierarchical control for different driving styles.

2.3.2   LOW-LEVEL CONTROLLER FOR DIFFERENT DRIVING STYLES

(1) *Controller for aggressive driving style:* Based on the sporty feature of aggressive driving style, the vehicle longitudinal control under this condition can be seen as an acceleration tracking problem, realizing the sporty feel in automated driving for passengers. Because of its ability to address nonlinearity and achieve good performance with fast response [38], a sliding-mode control (SMC) scheme is applied.

In designing the sliding-mode controller, the error term is defined as:

$$e = a - a_{ref}, \tag{2.16}$$

where a and a_{ref} are the actual and reference values of vehicle acceleration, respectively.

To guarantee zero steady error, an integral-type sliding surface S is chosen as:

$$S = \int edt. \tag{2.17}$$

One method for designing a control law that derives the system trajectories to the sliding surface is the Lyapunov direct method. The following Lyapunov function is used:

$$V = \frac{1}{2}SS. \tag{2.18}$$

To ensure the stability of the system, the derivative of the Lyapunov function should satisfy the following condition:

$$\dot{V} = S\dot{S} \leq 0. \tag{2.19}$$

Thus, if $\dot{S} = 0$, the above stability condition can be satisfied.

For the purpose of controller design, a control-oriented longitudinal vehicle model without considering wheel slip is used [35].

$$a = \frac{1}{mr} i_g T_m - fg - \frac{1}{2m} C_D A \rho v^2, \tag{2.20}$$

where r is the nominal radius, C_D is the coefficient of air resistance, A is the frontal area, ρ is the air density, f is the friction drag coefficient, and g is the gravitational acceleration.

Then, substituting Equations (2.16) and (2.20) into Equation (2.17), when $\dot{S} = 0$, the SMC control law can be derived as:

$$T_{m,ref} = \frac{mr}{i_g} \left(a_{ref} + fg + \frac{C_D A \rho v^2}{2m} - k_{SMC} \text{sgn}(S) \right), \tag{2.21}$$

where k_{SMC} is the positive gain of the SMC controller and $\text{sgn}(S)$ is the sign function defined as:

$$\text{sgn}(S) = \begin{cases} 1, & S > 0 \\ 0, & S = 0 \\ -1, & S < 0. \end{cases} \tag{2.22}$$

Remark 2.1   It is well known that in the standard SMC, the discontinuous sign function, $\text{sgn}(S)$, may cause chattering when the state trajectories are approaching the sliding surfaces. To avoid this phenomenon, the discontinuous term in Equation (2.21) could be replaced by a continuous function S, removing the chatter from the control input [39], as shown in Equation (2.23):

$$T_{m,ref} = \frac{mr}{i_g} \left(a_{ref} + fg + \frac{C_D A \rho v^2}{2m} - k_{SMC} S \right). \tag{2.23}$$

(2) *Controller for moderate driving style:* The moderate driving style features a balanced performance in vehicle dynamics and ride comfort. To this end, the low-level plant controller uses a combined feed-forward and feed-back structure, to actively damp powertrain torsional vibrations, thus mitigating the longitudinal jerk and enhancing drivability:

$$T_{m,ref} = T_{ff} + T_{fb}, \tag{2.24}$$

where T_{ff} is the feed-forward input term required for tracking and T_{fb} is the feedback component designed to reduce the control error.

Based on the control objective, the feed-forward term can be determined by the target motor torque $T_{m,tgt}$, which is calculated using the reference acceleration:

$$T_{ff} = T_{m,tgt}. \tag{2.25}$$

For the feedback term, a linear proportional-integral (PI) controller is adopted to damp the torsional oscillation:

$$T_{fb} = \left(K_P + K_I \int dt \right) \cdot e' \tag{2.26}$$

$$e' = T_{m,tgt} - 2T_{hs}/i_0 i_g, \tag{2.27}$$

where the feedback gains K_P and K_I are tuning parameters of the PI controller, and e' is the tracking error.

(3) *Controller for conservative driving style:* Since the conservative drivers usually care more about energy efficiency and smooth driving feel by carefully operating the brake and acceleration pedals, the low-level plant controller adopts the same combined feed-forward and feed-back architecture as the moderate one to ensure vehicle drivability.

2.4   DRIVING-STYLE-BASED PERFORMANCE EXPLORATION AND PARAMETER OPTIMIZATION

2.4.1   DESIGN SPACE EXPLORATION

Based on the system constrains formulated, namely the requirements for vehicle speed, gradeability, and brake stability shown in Equations (2.3)–(2.6), the boundaries of the related physical plant parameters can be calculated, and the design space is then achieved.

2.4.2   PERFORMANCE EXPLORATION METHODOLOGY

In order to carry out multi-objective optimization under different driving styles, the impacts of related parameters on the performance indicators should be explored. To do so, the following exploration algorithm is proposed (see Table 2.3).

As shown in Table 2.3, assuming that, within the Parameter Library ξ, there are several parameters, namely P_1, P_2, ..., P_i, C_1, C_2, ..., C_j, deciding one *Performance*. $P_1, P_2, \ldots, P_i$ represent parameters of the physical plant, while $C_1, C_2, \ldots, C_j$ indicate controller variables. Under pre-defined driving event E with valid design space, the selected vehicle *Performance* is simulated in the Simulink environment stepping each parameter with a suitably small step. After simulation-based global exploration, the *Best Performance K* with its corresponding value selections of the parameters can be attained.

2.4.3   DRIVING-STYLE-ORIENTED MULTI-OBJECTIVE OPTIMIZATION

(1) *Aggressive-driving-style based optimization:* This driving style requires to maximize vehicle dynamic performance first and foremost. However, a good performance in terms of energy efficiency is also expected to be guaranteed. Therefore, the trade-off between dynamic performance

Table 2.3: Algorithm for performance exploration

Algorithm 1: Performance Exploration
Input: Parameter Library $\{P_1, ..., P_i, C_1, ..., C_j\} \subseteq \xi$, Event E
Output: Best Performance Point K
function Global Exploration (ξ, E)
Performance $\leftarrow \{\}$; *Paras* $\leftarrow \{\}$;
while $p_1 \in P_1$ **do**
while $p_2 \in P_2$ **do**
$\vdots$
while $p_i \in P_i$ **do**
while $c_1 \in C_1$ **do**
while $c_2 \in C_2$ **do**
$\vdots$
while $c_j \in C_j$ **do**
Performance $\leftarrow$ *Simulation* $(E, P_1,..,P_i,C_1,...,C_j)$
end while
Paras $\leftarrow$ Performance (C_j);
$\vdots$
end while
Paras $\leftarrow$ Performance $(C_1, C_2, ..., C_j)$;
end while   ·
Paras $\leftarrow$ Performance $(P_i, C_1, C_2, ..., C_j)$;
$\vdots$
end while
Paras $\leftarrow$ Performance $(P_1, P_2, ..., P_i, C_1, C_2, ..., C_j)$;
K $\leftarrow$ Best Performance Point (Paras);
Return K, Paras
end function

Table 2.4: Weight selection for different styles

Driving Style	Weights			
	ω_1	ω_2	ω_3	ω_4
Aggressive	10	10	0	1
Moderate	10	0	10	1
Conservative	0	0	1	1

and energy efficiency is considered, with a much greater weight on the side of dynamic performance:

$$\{i_g, k_{SMC}, \beta\} = \arg \min_{-E_{reg}} (\omega_1 \cdot t_{acc} + \omega_2 \cdot t_{brk} + \omega_3 \cdot j - \omega_4 \cdot E_{reg}). \tag{2.28}$$

(2) *Moderate-driving-style based optimization:* In this case, the multi-objective optimization problem is set as a trade-off between dynamic performance and ride comfort:

$$\{i_g, K_P, K_I, \beta\} = \arg \min_{j} (\omega_1 \cdot t_{acc} + \omega_2 \cdot t_{brk} + \omega_3 \cdot j - \omega_4 \cdot E_{reg}). \tag{2.29}$$

(3) *Conservative-driving-style based optimization:* As mentioned before, under the conservative driving style, the drivers' behavior is usually mild with intentions of saving energy and ensuring comfort. Thus, in this mode, the trade-off elements are switched to ride comfort and energy efficiency:

$$\{i_g, K_P, K_I, \beta\} = \arg \min_{-E_{reg}} (\omega_1 \cdot t_{acc} + \omega_2 \cdot t_{brk} + \omega_3 \cdot j - \omega_4 \cdot E_{reg}). \tag{2.30}$$

For weighting selection, a much greater value would be put on the side of each featured performance under different driving styles, and the weight on non-considered performance is set as zero. The difference of the weights between featured and sub-featured performances are set to be an order of magnitude. The detailed set-up for the weightings under different driving styles is summarized in Table 2.4. The overall optimization flow and procedure are shown in Fig. 2.7.

2.5   OPTIMIZATION RESULTS AND ANALYSIS

Based on the proposed co-design method, the performance exploration and system optimization are carried out in MATLAB/Simulink. The simulations are implemented iteratively with developed models under defined driving events at each operating point (i.e., each selected value of plant and control parameters) for the three driving styles, generating multiple performances. The detailed results with each driving style are reported as follows.

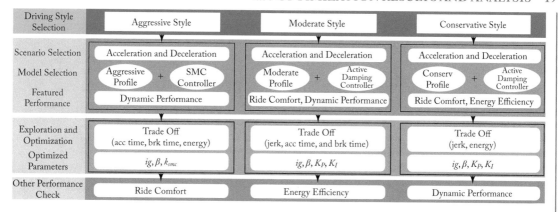

Figure 2.7: The proposed co-design optimization flow for the vehicle with three driving styles.

2.5.1 OPTIMIZATION RESULTS FOR THE AGGRESSIVE DRIVING STYLE

Since the optimization problem under the aggressive driving style is formulated as a trade-off between vehicle dynamic performance and energy efficiency with a much greater weight on the side of dynamic performance, during optimization the interactive effects of the values of the SMC gain, the gear ratio, and BFD on the dynamic performance of 0–50 km/h acceleration and regenerated braking energy are explored.

According to the exploration results shown in the subplots (a) and (b) of Fig. 2.8, the positive gain K of the SMC controller tends to be small, while the gear ratio prefers a larger value in favor of a better acceleration performance. For the regenerative braking performance, the variation of the gear ratio barely affects the overall regenerated energy, although BFD needs to select a smaller value to reach a higher efficiency according to the exploration results. This is due to the fact that more braking torque demand will be distributed to the front axle, which is the driven axle, indicating a larger proportion taken up by the regenerative braking among the overall braking torque.

2.5.2 OPTIMIZATION RESULTS OF THE MODERATE DRIVING STYLE

Based on the multiple optimization objectives under the moderate driving style, the trade-off between ride comfort and acceleration performance is considered. Taking the exploration scenario under a fixed value of the gear ratio at 8.3 as an example, and according to the results shown in the subplots (c) and (d) of Fig. 2.8, the selection of the gains in the linear PI controller for active damping has a great impact on the control performance of the vehicle jerk. With selection of K_P and K_I at 1.5 and 3.0, respectively, the maximum vehicle jerk during a 50–0 km/h deceleration process is over 10.0 m/s^3. While setting the two parameters to 0.5 and 2.0, the maximum jerk can be reduced to about 8.0 m/s^3, improving ride comfort effectively.

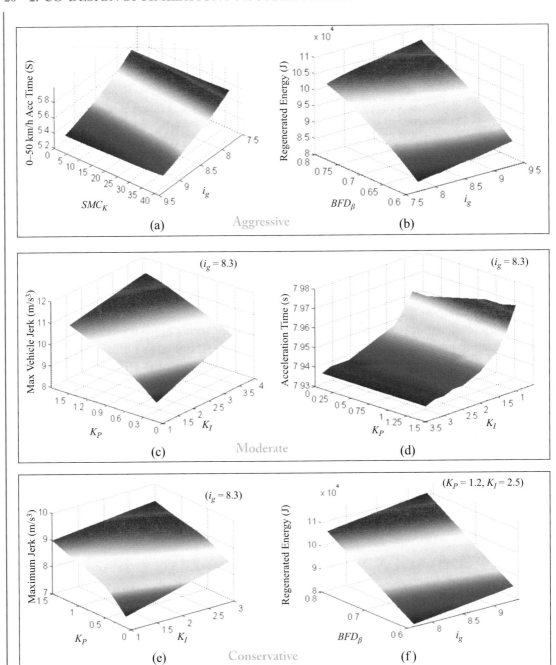

Figure 2.8: Performance exploration results of the three driving style.

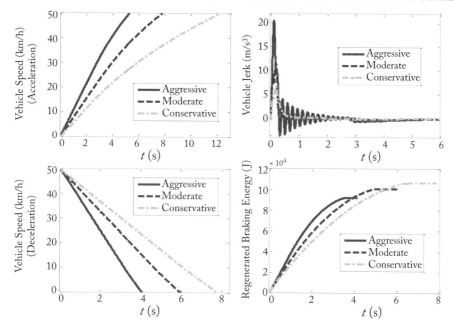

Figure 2.9: Optimized results for the vehicle under different driving styles.

However, the manipulation of the gains of the active damping controller has small influence on the acceleration performance, according to the exploration results.

2.5.3   OPTIMIZATION RESULTS OF THE CONSERVATIVE DRIVING STYLE

Since the controller structure of the conservative style is quite similar to the moderate one, the related parameters to be optimized (K_P, K_I, i_g, and β) are the same. However, because the optimization objectives are different under these two styles, the values of the parameters at the end of the optimization process can be far different, as the subplots (e) and (f) of Fig. 2.8 show.

2.5.4   COMPARISON AND DISCUSSION

A comparison of the above results is shown in Fig. 2.9. The aggressive style, which favors dynamic performance, dominates the acceleration and deceleration events among the three. The duration of the events of 0–50 km/h acceleration and 50–0 km/h deceleration under aggressive driving are 5.36 s and 4.16 s, respectively. The conservative style, which is in favor of ride comfort and energy efficiency, achieves the best performance in vibration reduction and regenerative braking. The maximum jerk under conservative driving is below 7 m/s³, which is around 1/3 of

Table 2.5: Optimized performance under different driving styles

Driving Style		Performance				
		t_{acc} /s	t_{brk} /s	j_{max} /m/s^3	E_{reg} /10^4 J	E_{ECE} /10^4 J
Aggressive	CPS-based	5.36	4.16	20.47	9.17	64.06
	w/o CPS	5.71	4.35	19.21	9.42	63.21
Moderate	CPS-based	7.88	6.04	11.52	10.04	60.19
	w/o CPS	9.26	6.35	11.91	9.49	62.06
Conservative	CPS-based	12.27	7.86	6.69	10.60	57.59
	w/o CPS	13.56	8.28	10.13	9.35	59.21

that in the aggressive driving. Finally, the moderate style, which sits in between the other two, achieves a good balance between dynamic performance, ride comfort, and energy efficiency.

To compare the energy efficiency at the vehicle level with designed control protocols and parameter selections during different driving styles, the standard ECE driving cycle is used. According to the test data in Table 2.5, the energy consumption of the automated electric vehicle under the conservative style is 575.9 kJ, which improves the efficiency by over 10%, compared to the energy used in aggressive driving.

Additionally, a comparison of the results between the CPS based optimization and the baseline is performed. According to the data listed in Table 2.5, the vehicle with CPS based optimization achieves better comprehensive performances in vehicle dynamics, ride comfort, and energy efficiency, thanks to the co-design of the plant and controller parameters. This demonstrates the advantages of the newly proposed method over the conventional one.

CHAPTER 3

State Estimation of Cyber-Physical Vehicle Systems

As we have mentioned, cyber physical systems, which are distributed, networked systems that fuse computational processes with the physical world exhibiting a multidisciplinary nature, have recently become a research focus [40–43]. As a typical application of CPS in green transportation, electric vehicles have been widely studied with different topics by researchers and engineers from academia, industry and governmental organizations [44–50]. In an electric vehicle (EV), the cyber world of control and communication, the physical plant of electric powertrain, the human driver, and the driving environment, are tightly coupled and dynamically interacted, determining the overall system's performance jointly [51]. These complex subsystems with multi-disciplinary interactions, strong uncertainties, and hard nonlinearities make the estimation, control, and optimization of electric vehicles very difficult [52]. Thus, there are still quite a number of fundamental issues and critical challenges varying in their importance from convenience to safety of EV remained open [53–56].

Among all those concerns in EV CPS, a key one is safety. Safety critical systems are those ones whose failure or malfunction may result in serious injury or severe damage to people, equipment, or environment [57]. As one of the most important safety critical systems in EV, the correct functioning of braking system is essential to the safe operation of the vehicle [58]. There are a variety of safety standards, control algorithms, and developed devices helping guarantee braking safety for current EVs. However, with increasing degrees of electrification, control authority and autonomy of CPVS, safety critical functions of braking system are also required to evolve to keep pace [59].

In the braking system of a passenger car, the braking torque is generated by the hydraulic pressure applied in the brake cylinder. Thus, the accurate measurement of the brake pressure through a pressure sensor is of great importance for various braking control functions and chassis stability logics. However, failures of the brake pressure measurement, which may be caused by software discrepancies or hardware problems, could result in vehicle's critical safety issues. Thus, high-precision estimation of brake pressure become a hot research area in CPVS design and control. Moreover, in order to handle the trade-offs between performance and cost, sensor-less observation is required. This makes the study of brake pressure estimation highly motivated.

Based on advanced theories and algorithms from the aspect of control engineering, observation methods of braking pressure for vehicles have been investigated by researchers worldwide. In [60], a recursive least square algorithm for estimation of brake cylinder pressure was proposed based on the pressure response characteristics of anti-lock braking system (ABS). In [61], an extended-kalman-filter-based estimation algorithm was developed considering hydraulic model and tyre dynamics. In [62], an algorithm for online observation of brake pressure was designed through a developed inverse model, and the algorithm was verified in the vehicle's electronic stability program. In [63], the models of brake pressure increase, decrease and hold are proposed, respectively, by using the experimental data. And the models can be used for fast online observation of hydraulic brake pressure. In [64], a brake pressure estimation algorithm was proposed for ABS considering the hydraulic fluid characteristics. In [65], the estimation algorithm was performed by calculating the volume of fluid flowing through the valve. The amount of fluid is a function of the pressure differential across the valve and the actuation time of the valve. Nevertheless, the existing research on brake pressure estimation was mainly investigated from the perspective of control engineering, and an approach with the probabilistic method, such as machine learning, has rarely been seen.

In this chapter, an Artificial-Neural-Network-based estimation method is studied for accurately observing the brake pressure of an electric passenger car. The main contribution of this work lies in the following aspects: (1) an Artificial Neural Network (ANN)-based machine learning framework is proposed to quantitatively estimate the brake pressure of an EV; (2) the proposed approach is implemented with experimental data obtained via vehicle testing, and compared with other methods; and (3) the proposed approaches has a great potential to achieve a sensorless design of the braking control system, removing the brake pressure sensor existing in the current products and largely reducing the cost of the system. Moreover, it also provides an additional redundancy for the safety-critical braking functions.

The rest of this chapter is organized as follows. Section 3.2 describes the high-level architecture of the proposed multilayer ANN for brake pressure estimation. Section 3.3 briefly introduces the standard backpropagation algorithm and illustrate the notations and basic concepts demanded in the Levenberg-Marquardt algorithm. Section 3.4 presents details of the application of the LMBP method to training the feed-forward neural networks. In Section 3.5, experiment implementations including feature selection, data collection, and preprocessing are presented. Section 3.5.1 reports the experimental results of the proposed brake pressure estimation algorithm including performance comparison to other approaches. Finally, conclusions are made in Section 3.5.3.

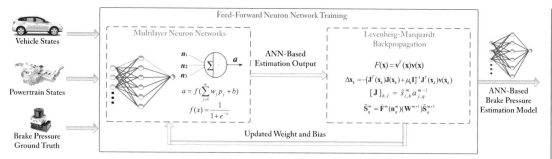

Figure 3.1: High-level architecture of the proposed brake pressure estimation algorithm based on multilayer Artificial Neural Networks.

3.1   MULTILAYER ARTIFICIAL NEURAL NETWORKS ARCHITECTURE

In order to achieve the objective of brake pressure estimation, multilayer artificial neural networks are first constructed with the input of vehicle and powertrain states. Details of the high-level system architecture and structure of the component are described in this section.

3.1.1   SYSTEM ARCHITECTURE

The system architecture with proposed methodology is shown in Fig. 3.1. The multilayer artificial neural network receives state variables of the vehicle and the electric powertrain system as inputs, and then yields the estimation of the brake pressure through the activation function. The Levenberg–Marquardt Backpropagation algorithm is then operated with the performance function, which is a function of the ANN-based estimation and the ground truth of brake pressure. The weight and bias variables are adjusted according to Levenberg–Marquardt method, and the backpropagation algorithm is used to calculate the Jacobian matrix of the performance function with respect to the weight and bias variables. With updated weights and biases, the ANN further estimates the brake pressure at the next time step. On the basis of the above iterative processes, the ANN-based brake pressure estimation model is well trained. The detailed method and algorithms are introduced in the following subsection.

3.1.2   MULTILAYER FEED-FORWARD NEURAL NETWORK

In this work, a multilayer feed-forward neural network is chosen to estimate brake pressure. A Feed-Forward Neural Network (FFNN) is composed of one input layer, one or more hidden layers and one output layer. ince a neural network with one hidden layer has the capability to handle most of the complex functions, in this work the FFNN with one hidden layer is constructed. Figure 3.2 shows the structure of a multilayer FFNN with one hidden layer.

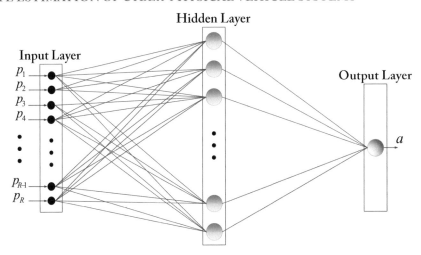

Figure 3.2: Structure of the multilayer feed-forward neural network.

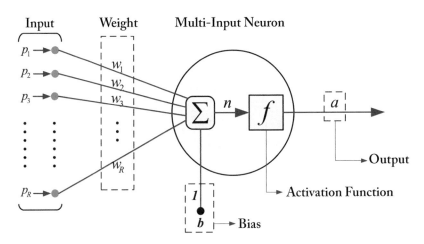

Figure 3.3: Structure of the multi-input neuron.

The basic element of a FFNN is the neuron, which is a logical-mathematical model that seeks to simulate the behavior and functions of a biological neuron [66]. Figure 3.3 shows the schematic structure of a neuron. Typically, a neuron has more than one input. The elements in the input vector $\mathbf{p} = [p_1, p_2, \ldots, p_R]$ are weighted by elements $w_1, w_2, \ldots, w_j$ of the weight matrix $\mathbf{W}$, respectively.

The neuron has a bias b, which is summed with the weighted inputs to form the net input n, which can be expressed by

$$n = \sum_{j=1}^{R} w_j p_j + b = \mathbf{W}_\mathbf{p} + b. \tag{3.1}$$

Then the net input n passes through an active function f, which generates the neuron output a:

$$a = f(n). \tag{3.2}$$

In this study, the log-sigmoid activation function is adopted. It can be given by the following expression:

$$f(x) = \frac{1}{1 + e^{-x}}. \tag{3.3}$$

Thus, the multi-input FFNN in Fig. 3.2 implements the following equation:

$$a^2 = f^2 \left(\sum_{i=1}^{S} w_{1,i}^2 f^1 \left(\sum_{j=1}^{R} w_{i,j}^1 p_j + b_i^1 \right) + b^2 \right), \tag{3.4}$$

where a^2 denotes the output of the overall networks. R is the number of inputs, S is the number of neurons in the hidden layer, and p_j indicates the jth input. f^1 and f^2 are the activation functions of the hidden layer and output layer, respectively. b_i^1 represents the bias of the ith neuron in the hidden layer, and b^2 is the bias of the neuron in the output layer. $w_{i,j}^1$ represents the weight connecting the jth input and the ith neuron of the hidden layer, and $w_{1,i}^2$ represents the weight connecting the ith source of the hidden layer to the output layer neuron.

3.2   STANDARD BACKPROPAGATION ALGORITHM

In order to train the established FFNN, the backpropagation algorithm can be utilized [67]. Considering a multilayer FFNN, such as the one with three-layer shown in Fig. 3.2, its operation can be described using the following equation:

$$\mathbf{a}^{m+1} = \mathbf{f}^{m+1} \left(\mathbf{W}^{m+1} \mathbf{a}^m + \mathbf{b}^{m+1} \right), \tag{3.5}$$

where $\mathbf{a}^m$ and $\mathbf{a}^{m+1}$ are the outputs of the mth and $(m + 1)$th layers of the networks, respectively. $\mathbf{b}^{m+1}$ is the bias vector of $(m + 1)$th layers of the networks. $m = 0, 1, \ldots, M - 1$, where M is the number of layers of the neural network. The neurons of the first layer obtain inputs:

$$\mathbf{a}^0 = \mathbf{p}. \tag{3.6}$$

Equation (3.6) provides the initial condition for Equation (3.5). The outputs of the neurons in the last layer can be seen as the overall networks' outputs:

$$\mathbf{a} = \mathbf{a}^M. \tag{3.7}$$

The task is to train the network with associations between a specified set of input-output pairs $\{(\mathbf{p}_1, \mathbf{t}_1), (\mathbf{p}_2, \mathbf{t}_2), \ldots, (\mathbf{p}_Q, \mathbf{t}_Q)\}$, where $\mathbf{p}_q$ is an input to the network, and $\mathbf{t}_q$ is the corresponding target output. As each input is applied to the network, the network output is compared to the target.

The backpropagation algorithm uses mean square error as the performance index, which is to be minimized by adjusting the network parameters, as shown in Equation (3.8):

$$F(\mathbf{x}) = E\left[\mathbf{e}^T \mathbf{e}\right] = E\left[(\mathbf{t} - \mathbf{a})^T (\mathbf{t} - \mathbf{a})\right], \tag{3.8}$$

where $\mathbf{x}$ is the vector matrix of network weights and biases. Using the approximate steepest descent rule, the performance index $F(\mathbf{x})$ can be approximated by

$$\hat{F}(\mathbf{x}) = (\mathbf{t}(k) - \mathbf{a}(k))^T (\mathbf{t}(k) - \mathbf{a}(k)) = \mathbf{e}^T(k)\mathbf{e}(k), \tag{3.9}$$

where the expectation of the squared error in Equation (3.8) has been replaced by the squared error at iteration step k.

The steepest descent algorithm for the approximate mean square error is

$$w_{i,j}^m(k+1) = w_{i,j}^m(k) - \alpha \frac{\partial \hat{F}}{\partial w_{i,j}^m} \tag{3.10}$$

$$b_i^m(k+1) = b_i^m(k) - \alpha \frac{\partial \hat{F}}{\partial b_i^m}, \tag{3.11}$$

where α is the learning rate.

Based on the chain rule, the derivatives in Equations (3.10) and (3.11) can be calculated as:

$$\frac{\partial \hat{F}}{\partial w_{i,j}^m} = \frac{\partial \hat{F}}{\partial n_i^m} \cdot \frac{\partial n_i^m}{\partial w_{i,j}^m}, \quad \frac{\partial \hat{F}}{\partial b_i^m} = \frac{\partial \hat{F}}{\partial n_i^m} \cdot \frac{\partial n_i^m}{\partial b_i^m}. \tag{3.12}$$

We now define s_i^m as the sensitivity of $\hat{F}$ to changes in the ith element of the net input at layer m:

$$s_i^m \equiv \frac{\partial \hat{F}}{\partial n_i^m}. \tag{3.13}$$

Using the defined sensitivity, then the derivatives in Equation (3.12) can be simplified as

$$\frac{\partial \hat{F}}{\partial w_{i,j}^m} = s_i^m a_j^{m-1} \tag{3.14}$$

$$\frac{\partial \hat{F}}{\partial b_i^m} = s_i^m. \tag{3.15}$$

Then the approximate steepest descent algorithm can be rewritten in matrix form as:

$$\mathbf{W}^m(k+1) = \mathbf{W}^m(k) - \alpha s^m \left(\mathbf{a}^{m-1}\right)^T \qquad (3.16)$$

$$\mathbf{b}^m(k+1) = \mathbf{b}^m(k) - \alpha s^m, \qquad (3.17)$$

where

$$s^m \equiv \frac{\partial \hat{F}}{\partial \mathbf{n}^m} = \left[\frac{\partial \hat{F}}{\partial n_1^m}, \frac{\partial \hat{F}}{\partial n_2^m}, \cdots, \frac{\partial \hat{F}}{\partial n_{S^m}^m}\right]^T. \qquad (3.18)$$

To derive the recurrence relationship for the sensitivities, the following Jacobian matrix is utilized:

$$\frac{\partial \mathbf{n}^{m+1}}{\partial \mathbf{n}^m} \equiv \begin{bmatrix} \dfrac{\partial n_1^{m+1}}{\partial n_1^m} & \dfrac{\partial n_1^{m+1}}{\partial n_2^m} & \cdots & \dfrac{\partial n_1^{m+1}}{\partial n_{S^m}^m} \\[2mm] \dfrac{\partial n_2^{m+1}}{\partial n_1^m} & \dfrac{\partial n_2^{m+1}}{\partial n_2^m} & \cdots & \dfrac{\partial n_2^{m+1}}{\partial n_{S^m}^m} \\[2mm] \vdots & \vdots & & \vdots \\[2mm] \dfrac{\partial n_{S^{m+1}}^{m+1}}{\partial n_1^m} & \dfrac{\partial n_{S^{m+1}}^{m+1}}{\partial n_2^m} & \cdots & \dfrac{\partial n_{S^{m+1}}^{m+1}}{\partial n_{S^m}^m} \end{bmatrix}. \qquad (3.19)$$

Consider the i, j element in the matrix:

$$\frac{\partial n_1^{m+1}}{\partial n_1^m} = w_{i,j}^{m+1} \frac{\partial a_j^m}{\partial n_j^m} = w_{i,j}^{m+1} \dot{f}^m \left(n_j^m\right). \qquad (3.20)$$

Thus, the Jacobian matrix can be rewritten as

$$\frac{\partial \mathbf{n}^{m+1}}{\partial \mathbf{n}^m} = \mathbf{W}^{m+1} \dot{\mathbf{F}}^m (\mathbf{n}^m), \qquad (3.21)$$

where

$$\dot{\mathbf{F}}^m (\mathbf{n}^m) = \begin{bmatrix} \dot{f}^m \left(n_1^m\right) & 0 & \cdots & 0 \\ 0 & \dot{f}^m \left(n_2^m\right) & & 0 \\ \vdots & \vdots & & \vdots \\ 0 & 0 & \cdots & \dot{f}^m \left(n_{S^m}^m\right) \end{bmatrix}. \qquad (3.22)$$

Then the recurrence relation for the sensitivity can be obtained by using the chain rule:

$$\begin{aligned} s^m = \frac{\partial \hat{F}}{\partial \mathbf{n}^m} &= \left(\frac{\partial \mathbf{n}^{m+1}}{\partial \mathbf{n}^m}\right)^T \frac{\partial \hat{F}}{\partial \mathbf{n}^{m+1}} \\ &= \dot{\mathbf{F}}^m (\mathbf{n}^m) \left(W^{m+1}\right)^T s^{m+1}. \end{aligned} \qquad (3.23)$$

This recurrence relation is initialized at the final layer as

$$s_i^M = \frac{\partial \hat{F}}{\partial n_i^M} = \frac{\partial ((\mathbf{t} - \mathbf{a})^T (\mathbf{t} - \mathbf{a}))}{\partial n_i^M} = \frac{\partial \sum_{j=1}^{S^M} (t_j - a_j)^2}{\partial n_i^M}$$
$$= -2 (t_i - a_i) \frac{\partial a_i}{\partial n_i^M} = -2 (t_i - a_i) \dot{f}^m \left(n_i^m \right).$$
(3.24)

Thus, the recurrence relation of the sensitivity matrix can be expressed as

$$\mathbf{s}^M = -2\dot{\mathbf{F}}^M (\mathbf{n}^M)(\mathbf{t} - \mathbf{a}).$$
(3.25)

The overall BP learning algorithm is now finalized and can be summarized as the following steps: (1) propagate the input forward through the network; (2) propagate the sensitivities backward through the network from the last layer to the first layer; and, finally (3) update the weights and biases using the approximate steepest descent rule.

3.3   LEVENBERG–MARQUARDT BACKPROPAGATION

While backpropagation is a steepest descent algorithm, the Levenberg–Marquardt algorithm is derived from Newton's method that was designed for minimizing functions that are sums of squares of nonlinear functions [68, 69].

Newton's method for optimizing a performance index $F(\mathbf{x})$ is

$$\mathbf{x}_{k+1} = \mathbf{x}_k - \mathbf{A}_k^{-1} \mathbf{g}_k$$
(3.26)

$$\mathbf{A}_k \equiv \nabla^2 F(\mathbf{x})|_{\mathbf{x}=\mathbf{x}_k}$$
(3.27)

$$\mathbf{g}_k \equiv \nabla F(\mathbf{x})|_{\mathbf{x}=\mathbf{x}_k},$$
(3.28)

where $\nabla^2 F(\mathbf{x})$ is the Hessian matrix and $\nabla F(\mathbf{x})$ is the gradient.

Assume that $F(\mathbf{x})$ is a sum of squares function:

$$F(\mathbf{x}) = \sum_{i=1}^{N} v_i^2(\mathbf{x}) = \mathbf{v}^T(\mathbf{x})\mathbf{v}(\mathbf{x})$$
(3.29)

then the gradient and Hessian matrix are

$$\nabla F(\mathbf{x}) = 2\mathbf{J}^T(\mathbf{x})\mathbf{v}(\mathbf{x})$$
(3.30)

$$\nabla^2 F(\mathbf{x}) = 2\mathbf{J}^T(\mathbf{x})\mathbf{J}(\mathbf{x}) + 2\mathbf{S}(\mathbf{x}),$$
(3.31)

where $\mathbf{J}(\mathbf{x})$ is the Jacobian matrix

$$\mathbf{J}(\mathbf{x}) = \begin{bmatrix} \dfrac{\partial v_1(\mathbf{x})}{\partial x_1} & \dfrac{\partial v_1(\mathbf{x})}{\partial x_2} & \cdots & \dfrac{\partial v_1(\mathbf{x})}{\partial x_n} \\ \dfrac{\partial v_2(\mathbf{x})}{\partial x_1} & \dfrac{\partial v_2(\mathbf{x})}{\partial x_2} & \cdots & \dfrac{\partial v_2(\mathbf{x})}{\partial x_n} \\ \vdots & \vdots & & \vdots \\ \dfrac{\partial v_N(\mathbf{x})}{\partial x_1} & \dfrac{\partial v_N(\mathbf{x})}{\partial x_2} & \cdots & \dfrac{\partial v_N(\mathbf{x})}{\partial x_n} \end{bmatrix} \tag{3.32}$$

and

$$\mathbf{S}(\mathbf{x}) = \sum_{i=1}^{N} v_i(\mathbf{x}) \nabla^2 v_i(\mathbf{x}). \tag{3.33}$$

If $\mathbf{S}(\mathbf{x})$ is assumed to be small then the Hessian matrix can be approximated as

$$\nabla^2 F(\mathbf{x}) \cong 2\mathbf{J}^T(\mathbf{x})\mathbf{J}(\mathbf{x}). \tag{3.34}$$

Substituting Equations (3.30) and (3.34) into Equation (3.26), we achieve the Gauss–Newton method as:

$$\Delta \mathbf{x}_k = -\left[\mathbf{J}^T(\mathbf{x}_k)\mathbf{J}(\mathbf{x}_k)\right]^{-1}\mathbf{J}^T(\mathbf{x}_k)\mathbf{v}(\mathbf{x}_k). \tag{3.35}$$

One problem with the Gauss–Newton method is that the matrix may not be invertible. This can be overcome by using the following modification to the approximate Hessian matrix:

$$\mathbf{G} = \mathbf{H} + \mu \mathbf{I}. \tag{3.36}$$

This leads to the Levenberg–Marquardt algorithm [70]:

$$\Delta \mathbf{x}_k = -\left[\mathbf{J}^T(\mathbf{x}_k)\mathbf{J}(\mathbf{x}_k) + \mu_k \mathbf{I}\right]^{-1}\mathbf{J}^T(\mathbf{x}_k)\mathbf{v}(\mathbf{x}_k). \tag{3.37}$$

Using this gradient direction, and recompute the approximated performance index. If a smaller value is yield, then the procedure is continued with the μ_k divided by some factor $\vartheta > 1$. If the value of the performance index is not reduced, then μ_k is multiplied by ϑ for the next iteration step.

The key step in this algorithm is the computation of the Jacobian matrix. The elements of the error vector and the parameter vector in the Jacobian matrix (3.32) can be expressed as

$$\mathbf{v}^T = [v_1 \; v_2 \; \dots \; v_N] = \left[e_{1,1} \; e_{2,1} \; \dots \; e_{S^M,1} \; e_{1,2} \; \dots \; e_{S^M,Q}\right] \tag{3.38}$$

$$\mathbf{x}^T = [x_1 \; x_2 \; \dots x_N] = \left[w_{1,1}^1 \; w_{1,2}^1 \; \dots w_{S^1,R}^1 \; b_1^1 \; \dots \; b_{S^1}^1 \; w_{1,1}^2 \; \dots \; b_{S^M}^M\right], \tag{3.39}$$

where the subscript N satisfies:

$$N = Q \times S^M \tag{3.40}$$

and the subscript n in the Jacobian matrix satisfies:

$$n = S^1(R+1) + S^2(S^1+1) + \cdots + S^M(S^{M-1}+1). \tag{3.41}$$

Making these substitutions into Equation (3.32), then the Jacobian matrix for multilayer network training can be expressed as

$$\mathbf{J}(\mathbf{x}) = \begin{bmatrix} \dfrac{\partial e_{1,1}}{\partial w_{1,1}^1} & \dfrac{\partial e_{1,1}}{\partial w_{1,2}^1} & \cdots & \dfrac{\partial e_{1,1}}{\partial w_{S^1,R}^1} & \dfrac{\partial e_{1,1}}{\partial b_1^1} & \cdots \\[2mm] \dfrac{\partial e_{2,1}}{\partial w_{1,1}^1} & \dfrac{\partial e_{2,1}}{\partial w_{1,2}^1} & \cdots & \dfrac{\partial e_{2,1}}{\partial w_{S^1,R}^1} & \dfrac{\partial e_{2,1}}{\partial b_1^1} & \cdots \\[2mm] \vdots & \vdots & & \vdots & \vdots & \\[2mm] \dfrac{\partial e_{S^M,1}}{\partial w_{1,1}^1} & \dfrac{\partial e_{S^M,1}}{\partial w_{1,2}^1} & \cdots & \dfrac{\partial e_{S^M,1}}{\partial w_{S^1,R}^1} & \dfrac{\partial e_{S^M,1}}{\partial b_1^1} & \cdots \\[2mm] \dfrac{\partial e_{1,2}}{\partial w_{1,1}^1} & \dfrac{\partial e_{1,2}}{\partial w_{1,2}^1} & \cdots & \dfrac{\partial e_{1,2}}{\partial w_{S^1,R}^1} & \dfrac{\partial e_{1,2}}{\partial b_1^1} & \cdots \\[2mm] \vdots & \vdots & & \vdots & \vdots & \end{bmatrix}. \tag{3.42}$$

In standard backpropagation algorithm, the terms in the Jacobian matrix is calculated as

$$\frac{\partial \hat{F}(\mathbf{x})}{\partial x_l} = \frac{\partial \mathbf{e}_q^T \mathbf{e}_q}{\partial x_l}. \tag{3.43}$$

For the elements of the Jacobian matrix, the terms can be calculated by

$$[\mathbf{J}]_{h,l} = \frac{\partial v_h}{\partial x_l} = \frac{\partial e_{k,q}}{\partial w_{i,j}}. \tag{3.44}$$

Thus, in this modified Levenberg–Marquardt algorithm, we compute the derivatives of the errors, instead of the derivatives of the squared errors as adopted in standard backpropagation.

Using the concept of sensitivities in the standard backpropagation process, here we define a new Marquardt sensitivity as

$$\tilde{s}_{i,h}^m \equiv \frac{\partial v_h}{\partial n_{i,q}^m} = \frac{\partial e_{k,q}}{\partial n_{i,q}^m}, \tag{3.45}$$

where $h = (q-1)S^M + k$.

Using the Marquardt sensitivity with backpropagation recurrence relationship, the elements of the Jacobian can be further calculated by

$$[\mathbf{J}]_{h,l} = \frac{\partial e_{k,q}}{\partial w_{i,j}^m} = \frac{\partial e_{k,q}}{\partial n_{i,q}^m} \frac{\partial n_{i,q}^m}{\partial w_{i,j}^m} = \tilde{s}_{i,h}^m a_{j,q}^{m-1} \tag{3.46}$$

if x_l is a bias,

$$[\mathbf{J}]_{h,l} = \frac{\partial e_{k,q}}{\partial b_i^m} = \frac{\partial e_{k,q}}{\partial n_{i,q}^m} \frac{\partial n_{i,q}^m}{\partial b_i^m} = \tilde{s}_{i,h}^m. \tag{3.47}$$

The Marquardt sensitivities can be computed using the same recurrence relations as the one used in the standard BP method, with one modification at the final layer. The Marquardt sensitivities at the last layer can be given by

$$\tilde{s}_{i,h}^M = \frac{\partial e_{k,q}}{\partial n_{i,q}^M} = \frac{\partial \left(t_{k,q} - a_{k,q}^M \right)}{\partial n_{i,q}^M} = -\frac{\partial a_{k,q}^M}{\partial n_{i,q}^M}$$

$$= \begin{cases} -\dot{f}^M \left(n_{1,q}^M \right) & \text{for } i = k \\ 0 & \text{for } i \neq k. \end{cases} \tag{3.48}$$

After applying the $\mathbf{p}_q$ to the network and computing the corresponding output $\mathbf{a}_q^M$, the LMBP algorithm can be initialized by

$$\tilde{\mathbf{S}}_q^M = -\dot{\mathbf{F}}^M \left(\mathbf{n}_q^M \right). \tag{3.49}$$

Each column of the matrix should be backpropagated through the network so as to generate one row of the Jacobian matrix. The columns can also be backpropagated together using

$$\tilde{\mathbf{S}}_q^m = \dot{\mathbf{F}}^m \left(\mathbf{n}_q^m \right) \left(\mathbf{W}^{m+1} \right) \tilde{\mathbf{S}}_q^{m+1}. \tag{3.50}$$

The entire Marquardt sensitivity matrices for the overall layers are then obtained by the following augmentation

$$\tilde{\mathbf{S}}^m = \left[\tilde{\mathbf{S}}_1^m | \tilde{\mathbf{S}}_2^m | \dots | \tilde{\mathbf{S}}_Q^m \right]. \tag{3.51}$$

3.4   EXPERIMENTAL TESTING AND DATA COLLECTION

In order to train the FFNN model with the LMBP algorithm proposed above and validate its effectiveness for brake pressure estimation, real vehicle driving data is needed. Thus, experiments using an electric passenger car are carried out on a chassis dynamometer. The testing vehicle together with the testing scenarios, selected feature vectors, data collection, and data pre-processing are described as follows.

3.4.1   TESTING VEHICLE AND SCENARIO

The experiment is implemented on a chassis dynamometer with an electric passenger car, as shown in Fig. 3.4a. The utilized electric vehicle is driven by a permanent-magnet synchronous motor, which is able to work in either driving or regenerating mode. The battery pack is connected to the electric motor via D.C. bus, releasing or absorbing power during driving and

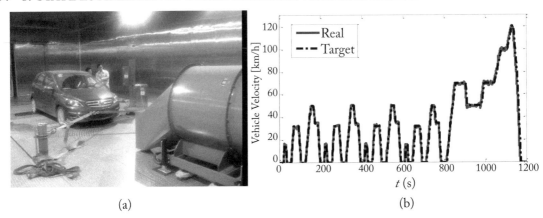

(a)   (b)

Figure 3.4: (a) The testing vehicle operating on a chassis dynamometer and (b) speed profile of the New European Drive Cycle (NEDC).

Table 3.1: Key parameters of the electric vehicle

Parameter	Value	Unit
Total vehicle mass	1360	kg
Wheel base	2.50	m
Frontal area	2.40	m^2
Nominal radius of tire	0.295	m
Coefficient of air resistance	0.32	—
Motor peak power	45	kW
Motor maximum torque	144	Nm
Motor maximum speed	9000	rpm
Battery voltage	326	V
Battery capacity	66	Ah

regenerative braking processes, respectively. Key parameters of the test vehicle are presented in Table 3.1.

To set up the testing scenario on a chassis dynamometer, standard driving cycles can be utilized. In this study, the New European Drive Cycle (NEDC) which consists of four repeated ECE-15 urban driving cycles and one Extra Urban Driving Cycle (EUDC) is adopted [71]. As Fig. 3.4b shows, the four successive ECE-15 driving cycles in the first section of the NEDC represent urban driving with low operating speed while the second section, i.e., the EUDC driving cycle, indicates a highway driving scenario with the vehicle speed up to 120 km/h.

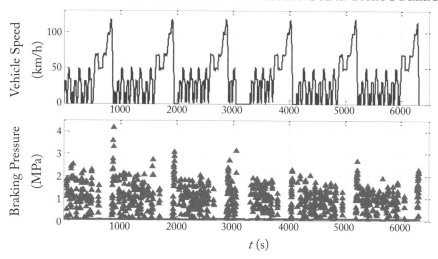

Figure 3.5: Collected data of the vehicle speed and corresponding brake pressure.

3.4.2   DATA COLLECTION AND PROCESSING

Vehicle data and powertrain states on CAN bus are collected with a sampling frequency of 100 Hz. Finally, experimental data of 6,327 s containing six NEDC driving cycles in total are recorded. The vehicle speed and brake pressure of the collected testing data during the four successive ECE-15 driving cycles are presented in Fig. 3.5.

In order to achieve a better training performance of the FFNN model with machine learning methods, the raw experimental data are smoothed at first using the following equation:

$$d_t = \frac{1}{N} \sum_n^N d_{tn},$$
(3.52)

where d_t is the value of a signal at time t, d_{tn} is the nth sampled value of signal d at time step t, and N is the total amount of samples within each second.

Then, in order to eliminate the effect brought by different units of signals utilized, the input signals are scaled to be in the range of 0–1.

3.4.3   FEATURE SELECTION AND MODEL TRAINING

In this work, the important vehicle and powertrain state variables are selected for the training of the multilayer ANN model for brake pressure estimation, while the real value of the brake pressure is utilized as a ground truth during the training process. When the electric vehicle is decelerating, the electric motor operates as a generator, recapturing vehicle's kinetic energy. During this period, the values of the motor and battery current change from positive to negative, indicating that the battery is charged by regenerative braking energy. Thus, apart from the vehicle

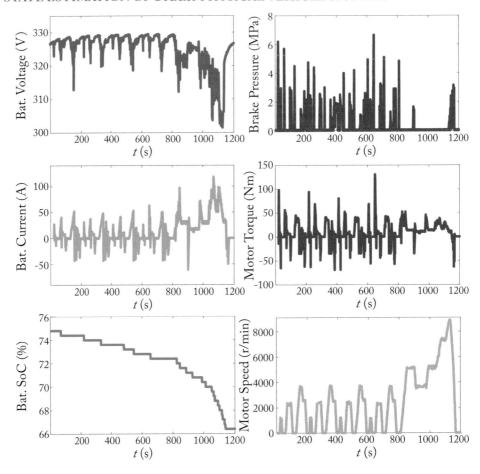

Figure 3.6: Experimental data of selected features during one driving cycle.

states, the signals of motor speed and torque, battery current and voltage, state of charge (SoC) are also chosen as features, i.e., the input vector of the FFNN. The data of some of the selected feature variables during one driving cycle are shown in Fig. 3.6.

Besides, statistical information, including the mean value, maximum value, and standard deviation (STD) of some of the vehicle states in the past few seconds are also adopted in this work. The features used for model training are listed below in Table 3.2.

After determining the feature vectors, the regression model of the FFNN is trained. To modulate and evaluate the model performance, the K-fold cross validation approach is adopted [72]. In this method, among the K folds divided, $(K - 1)$ ones are utilized to train the model, and the rest one fold is adopted for testing. Thus, the overall recorded data are divided into two sets, namely the training set and the testing one. The testing set, which is used for

Table 3.2: Selected features for FFNN model training

No.	Signal	Unit
1	Vehicle velocity	*km/h*
2	Mean value of velocity	*km/h*
3	STD of velocity	*km/h*
4	Maximum value of velocity	*km/h*
5	Vehicle acceleration	m/s^2
6	Motor speed	*rad/s*
7	Motor torque	*Nm*
8	Battery current	*A*
9	Battery voltage	*V*
10	Battery SoC	*%*
11	Gradient of battery voltage	*V/s*
12	Gradient of battery current	*A/s*

Table 3.3: Key parameters of the FFNN model

Parameter	Value
Maximum number of epochs to train	1000
Performance goal	0
Maximum validation failures	6
Minimum performance gradient	$1e^{-7}$
Initial μ	0.001
μ decrease factor	0.1
μ increase factor	10
Maximum μ	$1e^{10}$
Epochs between displays	25
Maximum time to train in seconds	Infinite

model validation, contains 1,400 samples chosen randomly from the raw data, and rest of the data are allocated to the training set. The final evaluation of the model performance is carried out based on the K test results. In this work, the value of K is set as 5. Then, with the 5-fold cross validation, the constructed FFNN is trained using the fast LMBP algorithm developed in Section 3.4. Some key parameters of the FFNN are illustrated in Table 3.3.

3.5   EXPERIMENT RESULTS AND DISCUSSIONS

In this section, results of the estimated ANN-based brake pressure with LMBP learning algorithm are presented and discussed. The algorithms are implemented in a computer with the MATLAB 2017a platform. The processor of the computer is an Intel Core i7-4710MQ CPU which supports 4 cores and 8 threads parallel computing, while the RAM equipped is a 32 G one. The time consuming for the FFNN training varies with the number of the hidden neurons selected. In this study, since the range of hidden neurons number is from 10–100, thus the training time for FFNN varies from 0.6–10 s, and the average training time cost for the FFNN with 70 neurons is 3.4 s.

3.5.1   RESULTS OF THE ANN-BASED BRAKING PRESSURE ESTIMATION

To quantitatively evaluate the estimation performance, two commonly used indices, namely the coefficient of determination R^2 and the root-mean-square-error (RMSE), are adopted. The definitions of the R^2 and RMSE are presented as follows. Suppose the reference data is $T = \{t_1 \ldots t_N\}$, and the predicted value is $Y = \{y_1 \ldots y_N\}$. Then R^2 can be calculated as:

$$R^2 = 1 - \frac{E_{res}}{E_{tot}} \tag{3.53}$$

$$E_{res} = \sum_{i}^{N} (t_i - y_i)^2 \tag{3.54}$$

$$E_{tot} = \sum_{i}^{N} \left(t_i - \bar{T}\right)^2, \tag{3.55}$$

where E_{res} is the residual sum of square, E_{tot} is the total sum of square, and $\bar{T}$ is the mean value of the reference data.

The RMSE can be obtained by:

$$RMSE = \sqrt{\frac{\sum_{i}^{N} (t_i - y_i)^2}{N}}. \tag{3.56}$$

First, the impact of the neuron number on the brake pressure estimation performance is analyzed. Considering the complexity of the problem, the estimation performance is tested under different number of neurons ranging from 10–100. According to Fig. 3.7, as the number of neurons changes, the estimation accuracy of the FFNN varies slightly. The best prediction performance is yield by FFNN with the number of neurons at 70.

Then, the linear regression performance of the trained model is investigated. Based on the linear regression result shown in Fig. 3.8, the test regression result R is of 0.96677, indicating

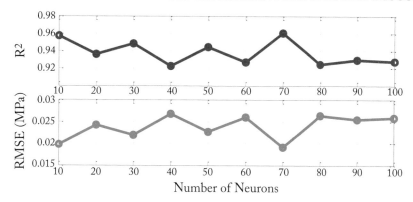

Figure 3.7: Estimation performance of FFNN with different number of neurons.

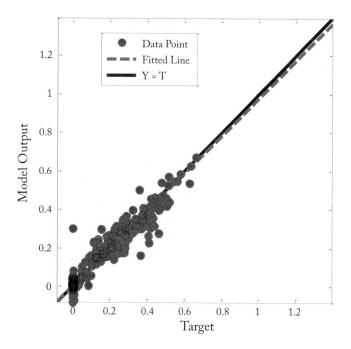

Figure 3.8: Regression performance of the FFNN model with 70 neurons.

the FFNN model with 70 neurons can accurately estimate the braking pressure through selected features.

Figure 3.9 shows the brake pressure estimation result in time domain. The x-axis presents the 1,400 samples of the testing data set, and the y-axis shows the estimation results of the scaled brake pressure. Since the input and output data for model training is scaled to the range of $[0, 1]$,

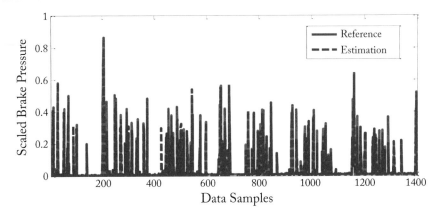

Figure 3.9: ANN-based braking pressure estimation results with 1,400 testing data points.

the model testing output is then falling within the range between 0 and 1 accordingly. Based on the results, the FFNN model achieves high-precision regression performance, and the *RMSE* is around 0.1 MPa, demonstrating the feasibility and effectiveness of the developed method.

3.5.2   IMPORTANCE ANALYSIS OF THE SELECTED FEATURES

Besides, the utilized feature variables are further investigated through analyzing the importance of predictors [73]. A larger value of the predictor importance indicates that the feature variable has a greater effect on the model output.

Figure 3.10 illustrates the estimation results of the predictor importance. Based on the results, the most important feature in the model is the battery current, followed by STD of velocity, vehicle velocity, and acceleration. Besides, the battery voltage, the gradients of the battery voltage and current also exert impacts on the model estimation performance.

3.5.3   COMPARISON OF ESTIMATION RESULTS WITH DIFFERENT LEARNING METHODS

The developed ANN-based approach is compared with other machine learning methods, including regression decision tree, Quadratic support vector machine (SVM), Gaussian process model, and regression Random Forest. These models are also trained and tested with the 5-fold cross validation method. Apart from R^2 and *RMSE*, other two evaluation parameters, i.e., the training time and the testing time, are also utilized to assess the performance of different models.

Detailed results of the comparison are shown in Table 3.4. According to the results, the single decision tree algorithm gives much shorter training time and a much faster testing speed in comparison to the other algorithms. In terms of real-time application, the regression decision tree could be a good candidate because of its simplicity and high computation efficiency. How-

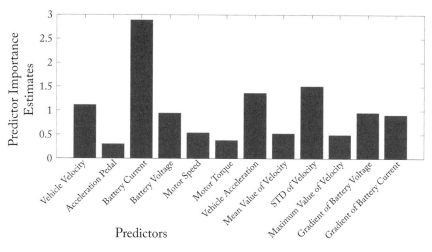

Figure 3.10: The predictor importance estimation results.

Table 3.4: Comparison of braking pressure estimation performance

Method	R^2	RMSE (MPa)	Training Time (s)	Testing Speed(obs/s)
Decision Tree	0.912	0.133	1.092	~240000
Quadratic SVM	0.867	0.188	141.93	~46000
Gaussian Process model	0.921	0.125	156.89	~8100
Random Forest	0.903	0.104	3.79	~36000
ANN	0.935	0.101	3.42	~82000

ever, with respect to the brake pressure estimation accuracy (both R^2 and *RMSE*), the developed ANN algorithm yields the best performance with acceptable training time and testing speed.

CHAPTER 4

Controller Design of Cyber-Physical Vehicle Systems

As a typical application of the safety-critical CPVS, the brake-by-wire system, which mechanically decouples the brake pedal with downstream brake circuits, features the flexibility of the braking circuit arrangement and high-precision pressure modulation. Because of the inherent pedal-decoupling, the BBW system is usually chosen to be the platform for implementing brake blending control of different forms of electrified vehicles [74–76]. Moreover, the possibility of fast and accurate modulation of mechatronic actuators is the foundation of many vehicle dynamics control systems and intelligent vehicles [77–80].

As the BBW system promises benefits including efficiency and precision, it has become a popular area, attracting increasingly interest from both industry and academia. Currently, there are two types of BBW system that have been developed: the electrohydraulic brake (EHB) and the electromechanical brake (EMB). In EHB system, the brake pressure is established by a hydraulic pump which is driven by an electric motor. In addition, high pressure accumulators and proportional valves are always equipped to achieve accurate hydraulic pressure control and noise reduction [81–83]. While the EMB solution removes all the hydraulic components, instead, it adopts an electric motor as actuator to provide braking torque, largely reducing the system complexity [84–86]. Although existing BBW systems can achieve continuous modulation of the braking torque, the high cost of proportional valves and the complexity of high pressure accumulators and actuators of EMB limit their usage. This limitation motivated us to develop a new type of BBW system that simply adds conventional actuators and sensors to available hydraulic braking layouts, realizing an overall control performance equivalent to that of an existing EHB.

In conventional hydraulic braking systems, the low-cost switching valves are widely utilized. By applying pulse-width modulation (PWM) control, it can achieve an acceptable control accuracy. However, the inherent discrete behavior of on/off valves causes not only the degradation of the pressure modulation accuracy but also an increase in the high frequency noise that passengers find most intolerable. Thus, several comprehensive research studies have been conducted into the areas of parameter design and control methods for on/off valves to address the above problems. In [87], valve parameter optimizations were conducted to enhance the dynamic performance of hydraulic valves. The relationship between dynamic performance and valve geometric parameters were analyzed. In [88], in order to compensate the control precision degradation caused by the discrete nature of actuator dynamics, a sliding mode control based method was proposed. Using the similar idea, a modified PWM-based algorithm was proposed to enhance the responsiveness of on-off valves [89]. A learning-based method was implemented to guarantee the effectiveness of the proposed approach in the face of various external disturbances. In [90], to mimic the performance of the proportional valve, a PWM-based modulation approach with a high frequency ranging from 2–5 kHz was developed for on/off valves. In [91], an observer was designed for position estimation of valve core; together with the sliding mode control technique which is similar to [17], the pressure modulation accuracy was improved. Nevertheless, because of the discrete nonlinearity of on-off valves, the existing control methods mentioned above can hardly achieve a desirable linear performance as a proportional valve behaves. Moreover, the noise reduction has been rarely considered in the existing studies.

A novel quasi-BBW system and a linear control method for on/off valves were proposed in [75, 92–95]. However, the proposed system can only decouple the mechanical connection between the braking pedal and the hydraulic brake circuits in some specific cases. Moreover, the proposed linear control method is an open loop method, which is sensitive to external disturbances. Motivated by the system and the control method in [75, 92–95], in this chapter, a novel BBW system is designed simply by adding conventional actuators and sensors to the usual hydraulic layouts. To improve the pressure modulation accuracy of the hydraulic control system and reduce noise, two pressure modulation methods, namely, the hydraulic pump-based pressure modulation (HPBPM) and the closed-loop pressure-difference-limiting (CLPDL) modulation, are proposed and compared. The HPBPM is developed to increase the hydraulic braking modulation precision by coordinating the hydraulic pump and corresponding solenoid valves, while CLPDL leverages on the valve current control technology. The main novelty and contribution of this work are as follows: (1) the novel BBW system is developed simply by adding conventional actuators and sensors to the conventional hydraulic component layout that is simpler than the system in [96]; (2) two pressure modulation methods of the newly developed BBW system, namely, HPBPM and CLPDL, are proposed, and the control performance and noise were are compared via HiL validation; and (3) the CLPDL control method is applied to a brake blending control strategy of EVs, validating the effectiveness and feasibility of the CLPDL method in actual implementation.

The remaining parts of this work are organized as follows. In Section 4.2, the system and working principle of the novel BBW is described. In Section 4.3, the HPBPM for the BBW system is proposed. In Section 4.4, the linear correlation between the pressure difference across the valve and applied coil current is analyzed, and the CLPDL method of the BBW system is developed. In Section 4.4.1, the performances with respect to tracking control and noise level of the system under HPBPM and CLPDL methods are compared; the effectiveness of the CLPDL approach is demonstrated by a normal regenerative braking scenario. In Section 4.4.2, the study's conclusions are provided.

4.1 DESCRIPTION OF THE NEWLY PROPOSED BBW SYSTEM

A novel BBW system, which can be implemented on all types of vehicles, including traditional internal combustion engine (ICE) vehicles and different forms of electrified and intelligent ones, is schematically depicted in Fig. 4.1. Unlike the standard BBW system—for example, Toyota's electronically controlled brake (ECB), [96]—the high pressure accumulator and linear valves are not needed in the proposed system; all of the valves in the proposed system are normal on/off valves. Only two relief valves (RV1, RV2) are added, unlike the standard electronic stability program (ESP) systems [97], the differences between proposed BBW system and conventional BBW system have been highlighted in Fig. 4.1. The pressures of the master cylinder and each wheel cylinder are measured by the corresponding pressure sensors.

To achieve good tracking performance of the target pressure, the wheel cylinder pressure should be modulated precisely. The state machine for wheel cylinder pressure modulation is illustrated in Fig. 4.2. The wheel cylinder pressure is threshold controlled and has three modes, including pressure increase, pressure hold and pressure decrease. These modes can shift from one to another by judging the logic thresholds, which are the various differences between the actual and target wheel cylinder pressures. The operation principle of the developed BBW system is explained in details below.

During the normal brake mode, the four normally open isolation valves are activated and closed. Then, the master cylinder pressure, which represents driver's deceleration demand, is physically isolated from the brake pressures of the downstream wheel cylinders. The braking fluid in the master cylinder is directly passed into the pedal stroke simulator, ensuring good brake pedaling feeling. As all of the wheel cylinders have the same braking circuit layout, we take the example of the front right wheel to explain the pressure modulation process.

During the pressure increase process, two hydraulic pumps are driven by the hydraulic pump-motor, which is controlled by PWM signals. Brake fluid is pumped directly from the tank instead of the master cylinder, leaving the brake pedaling feel unaffected. The opening of the normally closed relief valve RV1 is controlled to keep the upper stream pressure of the inlet valve IV1 constant or at the upper limit, depending on the relief valve (RV1) control method. By adjusting the opening of the inlet valve IV1, the wheel cylinder pressure can be modulated. The precise pressure increase is achieved by the coordinated control of the hydraulic pump, relief valve, and inlet valve. The detailed control algorithms for relief and inlet valves will be investigated in Sections 4.3 and 4.4.

During the pressure decrease process, the hydraulic pumps stop and the relief valve RV1 is fully open. At the same time, the IV1 is fully closed, and the outlet valve OV1 is controlled with PWM signals, which modulates the decreasing ratio of the brake pressure in wheel cylinder. The hydraulic fluid flows from the wheel cylinder to the tank directly.

During the pressure hold process, both inlet valve IV1 and outlet valve OV1 are fully closed. The pump operates at a low speed, and the relief valve RV1 is opened to a certain level.

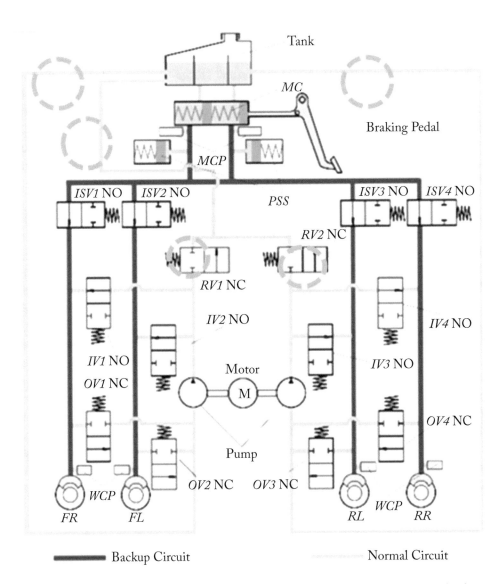

Figure 4.1: Diagram of the structure of the proposed BBW system. *MC*: master cylinder; *MCP*: master cylinder pressure sensor; *PSS*: braking pedal stroke simulator; *ISV*: isolation valve; *RV*: relief valve; *IV*: inlet valve; *OV*: outlet valve; *WCP*: wheel cylinder pressure sensor; *FR*: front right wheel; *FL*: front left wheel; *RL*: rear left wheel; *RR*: rear right wheel; *NO*: normally open valve; *NC*: normally closed valve.

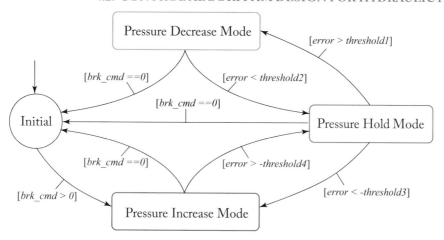

Figure 4.2: State machine of the wheel cylinder pressure modulation (error: difference between the actual and the target wheel cylinder pressures; brk_cmd: driver braking intention that can be interpreted from the master cylinder pressure).

This property keeps the pressure of the entire braking circuit above a certain level, which ensures responsiveness for the next pressure increase process.

As the pressure of the four wheel cylinders can be controlled independently, the proposed BBW system can also be used for other control functions, including anti-lock braking, traction control, and stability control program. In addition, this novel design can be installed on vehicles with any type of braking circuit arrangement. Note that when the hydraulic pump fails, as in a conventional vehicle, a backup circuit ensures the pressure build up by drivers; this is highlighted in Fig. 4.1.

4.2   CONTROL ALGORITHM DESIGN FOR HYDRAULIC PUMP-BASED PRESSURE MODULATION

As mentioned in the introductory section, the following two performances must be guaranteed: first, the tracking performance of the target wheel cylinder pressure; and second, low noise during BBW system operation. Thus, the control algorithms for the on/off valves and hydraulic pump are worth investigating. Two different control schemes have been implemented; these schemes are presented in separate sections.

The overall controller configuration for pressure modulation of the two different scheme is illustrated in Fig. 4.3. In the first scheme, we aimed at achieving the control goals by employing the conventional PWM control technique for inlet valves, and a newly proposed coil current control technique is implemented in the second scheme. The first scheme is introduced as follows.

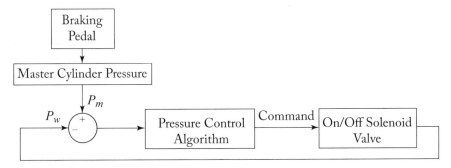

Figure 4.3: Overall controller configuration of pressure modulation.

The noise generated by the BBW system mainly originates from two parts: the on/off valve and the hydraulic pump. Compared to the valve generated noise, which is mostly centralized in the high frequency range, the hydraulic pump contributes more to low frequency noise. However, the high frequency noise is most perceivable and intolerable for passengers, especially when the opening of an on/off value is being modulated frequently. Thus, an intuitive method to reduce the annoying high frequency noise is to keep the PWM command of the inlet valve constant. The tracking performance of the brake pressure in a wheel cylinder is guaranteed by modulation of the rotational speed of the pump-motor. To illustrate the main idea of the HPBPM method, a simple but valid spring and position model [90–95] is used. The dynamics of the wheel cylinder pressure p_w and the displacement of the spring x_w of the cylinder can be given by:

$$dp_w = \frac{k_w}{\pi r_w^2} \cdot \frac{\sigma_{PWM}}{100} dx_w \qquad (4.1)$$

$$dx_w = \frac{Q}{\pi r_w^2} dt, \qquad (4.2)$$

where k_w represents the stiffness coefficient of the spring, r_w is the of the piston in the brake cylinder, σ_{PWM} (which varies within 0–100) represents the value of the PWM signal for inlet valve, and Q is the flow rate of the fluid passing through the valve. According to the working principle of the hydraulic pump, the flow rate Q is expressed as

$$Q = 2\pi r_{pump}^2 \eta_{pump} e_{pump} \omega_{hyd\_motor}, \qquad (4.3)$$

where r_{pump}, η_{pump}, e_{pump}, and ω_{hyd_motor} are the radius of the piston, efficiency, eccentricity, and rotational speed of the pump-motor, respectively. By substituting Equations (4.2) and (4.3) in (4.1), the wheel cylinder pressure p_w can be rewritten as

$$\dot{p}_w = \frac{2k_w r_{pump}^2 e_{pump} \eta_{pump}}{\pi r_w^4} \cdot \frac{\sigma_{PWM}}{100} \cdot \omega_{hyd\_motor}. \qquad (4.4)$$

It can be seen from Equation (4.4) that when the PWM command value of the inlet valve is constant, and the dynamics of the pressure is directly decided by the rotational velocity of the pump-motor. Based on the property of PWM control, a suitable PWM command value for the inlet valve should be chosen based on the following criteria. Regarding the first criterion, the valve generated noise becomes most perceivable and intolerable when the PWM command value is approximately 50, at which the impact between valve coil and seat is most dramatic.

Thus, less noise will be generated when the inlet valve PWM command chosen is closer to 0 or 100. Regarding the second criterion, because of the discontinuous property of the hydraulic pump, pressure fluctuations will be generated. These fluctuations can be reduced by opening the inlet valve up to a small level. In other words, the PWM command is supposed to be close to 100 for a normally open inlet valve.

Figure 4.4a shows the relationship between the variation rate of the brake pressure in wheel cylinder and the value of the PWM signal of the pump-motor. In both experiments, the inlet valve PWM value is set to 80, at which the noise is acceptable, and the pressure fluctuation generated by the hydraulic pump is adequately suppressed. A linear relationship between the variation rate of the wheel cylinder pressure and the value of the PWM signal for the pump-motor is observed, as shown in Fig. 4.4b. This result validates the correctness of Equation (4.4).

To improve the tracking performance and robustness to external disturbances, a closed-loop control algorithm for the hydraulic pump-motor is developed, as illustrated in Fig. 4.5. The feed-forward lookup map is derived by experiments, as shown in Fig. 4.4. Because the wheel cylinder pressure can be obtained by the pressure sensor, feedback control is utilized to offset the wheel cylinder pressure tracking error. Note that the relief valve, which is controlled inactively, plays a role only of safety outlet valves in the HPBPM control method. When the upper stream pressure of the inlet valve exceeds 15 MPa, the relief valve will be pushed open to release the excess pressure. At other times, the upper stream pressure of the inlet valve is equal to the outlet pressure of the hydraulic pump.

4.3   CONTROL ALGORITHM DESIGN FOR CLOSED-LOOP PRESSURE-DIFFERENCE-LIMITING MODULATION

The coil current-based valve control method, which is different from the control method discussed in the last section, is explored in this section. A closed-loop modulation method is developed based on the previously discussed linear property for on/off valves.

4.3.1   LINEAR MODULATION OF ON/OFF VALVE

The relative position of a normally open valve in closed-state is illustrated in Fig. 4.6. Taking the lowest point which valve core can reach as the origin, the local coordinate system of the valves can be set up [45].

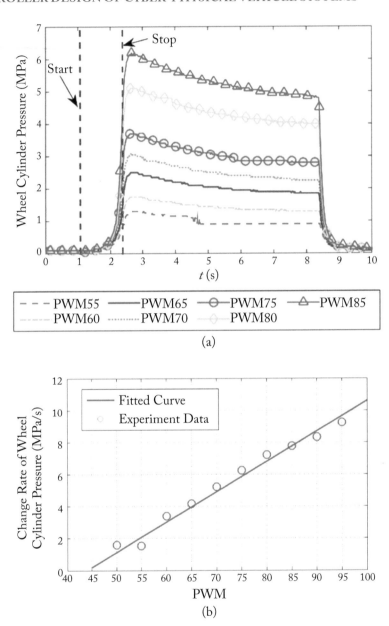

Figure 4.4: (a) Relationship between the wheel cylinder pressure and the value of the PWM signal of the pump-motor (with the PWM value of the inlet valve at 80); and (b) relationship between the wheel cylinder pressure and different PWM command values of the pump-motor (with the PWM value of the inlet valve at 80).

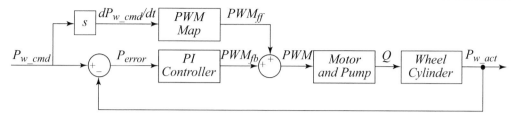

Figure 4.5: Structural diagram of the PWM control algorithm for wheel cylinder pressure.

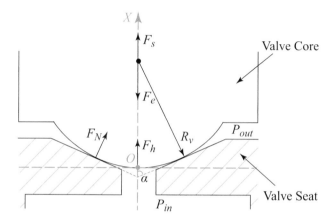

Figure 4.6: Diagram of the coordinate system of the inlet valves.

Suppose the valve core comes in contact with the valve seat after being energized, the axial forces applied on the valve can be represented as

$$-F_e + F_s + F_h + F_N \sin \alpha = 0, \tag{4.5}$$

where F_e, F_s, F_h, F_N, and α are the electromagnetic force, spring, hydraulic force, supportive force, and valve seat cone angle, respectively [45].

The valve can be defined to be in a critical balanced position when the following two conditions are satisfied: first, the valve core is in contact with the valve seat; second, the contact force between valve core and valve seat is zero, i.e., $F_N = 0$. In this state, the valve core is just about to leave the valve seat and the force balance equation in (4.5) is simplified as

$$-F_e + F_s + F_h = 0. \tag{4.6}$$

In Equation (4.5), the only term that can be actively controlled is the electromagnetic force F_e. Thus, to realize the critical balanced state, the electromagnetic force should be regulated in a way that makes $F_N = 0$. The electromagnetic force can be expressed as

$$F_e = \frac{(NI)^2}{2R_g l}, \tag{4.7}$$

where I and N are the applied current and turns number of the coil, respectively. l and R_g are the length and magnetic reluctance of air gap [45]. Then, we linearize Equation (4.7) at a fixed operation point and reform it as:

$$
\begin{aligned}
F_e &= \left.\frac{\partial F_e}{\partial I}\right|_{I=I(t)} \cdot I(t) + \left.\frac{\partial F_e}{\partial x_v}\right|_{x_v=x_v(t)} \cdot x_v(t) \\
&= K_i I(t) + K_{x_v} x_v(t),
\end{aligned} \tag{4.8}
$$

where x_v is the displacement of the valve core, K_i and K_{x_v} are the first-order current-force and displacement-force coefficient.

Note that the displacement of valve core is $x_v = 0$ when the critical balanced position is reached. Then, Equation (4.8) can be simplified as

$$F_e = K_i I(t). \tag{4.9}$$

Based on the definition of the OX coordinate in Fig. 4.6, the spring force of a normally open valve is expressed as

$$F_s = k_s (x_0 + x_m - x_v), \tag{4.10}$$

where x_0, x_m, and k_s present the pre-tension displacement, maximum displacement and the stiffness of the return spring, respectively. Besides, the spring force can be simplified in this state as

$$F_s = k_s (x_0 + x_m). \tag{4.11}$$

When the valve core reaches the fully closed position, the F_h is decided by the pressure drop Δp between valve inlet and outlet and A_s, which is the surface area exposed to the fluid in axial direction. F_h, A_s, and Δp are calculated as follows [45]:

$$F_h = \Delta p \cdot A_s \tag{4.12}$$

$$A_s = \pi R_v^2 (\cos \alpha)^2 \tag{4.13}$$

$$\Delta p = p_{in} - p_{out}, \tag{4.14}$$

where R_v is the spherical radius in the valve core.

Substituting Equations (4.9), (4.11), and (4.12) into Equation (4.6), we can derive a linear correlation between coil current and the pressure difference at valve's critical balanced position [45]:

$$\Delta p = \frac{K_i}{\pi R_v^2 (\cos \alpha)^2} I - \frac{k_s (x_0 + x_m)}{\pi R_v^2 (\cos \alpha)^2}. \tag{4.15}$$

Substituting Equation (4.14) in Equation (4.15), we can obtain the linear modulation property of valve outlet pressure by adjusting the coil current,

$$p_{out} = p_{in} - \frac{K_i}{\pi R_v^2 (\cos \alpha)^2} I + \frac{k_s (x_0 + x_m)}{\pi R_v^2 (\cos \alpha)^2}. \tag{4.16}$$

If the input pressure p_{in} can be measured or estimated, we will be able to address the output pressure tracking problem by calculating the coil current command value based on Equation (4.16) as has been done in [45]. However, the following two problems must be addressed to achieve better tracking performance of the output pressure. First, as mentioned in the introductory section, the control algorithm, which directly implements Equation (4.16) in an open-loop manner as done in [45], is not robust enough to disturbances. Second, unlike the case in [45], the value of the input pressure p_{in} can be acquired accurately by using the pressure sensor. However, the pressure of the valve input port cannot be measured in the proposed BBW system. Thus, a closed-loop tracking control algorithm is required to be designed for the proposed system, ensuring wheel cylinder pressure control accuracy and low noise generation.

4.3.2   CLOSED-LOOP PRESSURE-DIFFERENCE-LIMITING CONTROL

The structure of the CLPDL control for hydraulic pressure modulation is shown in Fig. 4.7. To improve the robustness and accuracy of the control algorithm, a feed-forward and feed-back scheme is employed. The current map which reflects the linear characteristics as Equation (4.15) shows is derived from the off-line experiments. A simple PI controller is utilized to eliminate external disturbances, such as temperature change and input pressure fluctuation.

p_{w_act} can be obtained by the hydraulic pressure sensor installed in the wheel cylinder. However, there is no sensor installed for measuring the inlet pressure p_{input}. Motivated by the methods adopted in conventional BBW systems, the ideal p_{input} is to be controlled to a stable value. To achieve a stable upstream pressure for the inlet valve, a coordinate control strategy for the pump-motor and relief valve is developed. Unlike the method in Section 4.3, the relief valve is controlled actively. Note that the relief valve is normally closed; thus, the relation shown in Equation (4.16) needs to be rewritten as follows:

$$p_{out\_relief} = p_{in\_relief} - \frac{k_{s\_relief} (x_0 + x_m)}{\pi R_{v\_relief}^2 \left(\cos \alpha_{relief} \right)^2} + \frac{K_{i\_relief}}{\pi R_{v\_relief}^2 \left(\cos \alpha_{relief} \right)^2} I, \tag{4.17}$$

where p_{in_relief} and p_{out_relief} are the input and output pressures of the relief valve, respectively. As indicated in Fig. 4.1, the output port of the relief valve is directly connected to the tank, where the pressure is zero; thus, $p_{out\_relief} = 0$. The upstream pressure for the inlet valve is equal to p_{in_relief}; therefore, Equation (4.17) can be simplified as follows:

$$p_{in\_relief} = \frac{k_{s\_relief} (x_0 + x_m)}{\pi R_{v\_relief}^2 \left(\cos \alpha_{relief} \right)^2} - \frac{K_{i\_relief}}{\pi R_{v\_relief}^2 \left(\cos \alpha_{relief} \right)^2} I. \tag{4.18}$$

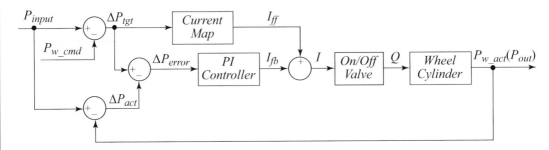

Figure 4.7: Structural diagram of the CLPDL control algorithm for wheel cylinder pressure modulation.

Based on the analysis above, by setting constant command values for the relief valve and pump-motor, the stable upstream pressure for the inlet valve can be derived.

According to the working principle of the pressure-difference-limiting control, the upstream pressure for the input valve should be greater than or equal to the current driver braking intensity. In the developed BBW system, accurate modulation of the coil current is achieved by the current control chip, whose best working region is from 100–600 mA. Thus, the optimal value of the pressure difference is set within the range from 0–4 MPa. A switched coordinated control strategy for relief valve and pump-motor has been designed, as shown in Fig. 4.8. The pressure increase mode has two sub-phases, denoted as low and high braking intensity mode, which can realize the different stable upper stream pressure for inlet valves. This property ensures that the current control chip works within its optimal region and that there is enough up stream pressure for the inlet valves. Hydraulic pump only works in the pressure hold and increase modes. Because no noise is generated when the valve is controlled by the current method, the noise is significantly reduced with CLPDL control.

Note that the current command value for relief valves RV_cm1, RV_cmd2, and RV_cmd2, are different for the brake cylinders in front and rear wheels. However, for the same wheel cylinder, the relative relationship among the three values is the same, which given by is $0 < RV\_cmd3 < RV\_cm1 < RV\_cmd2$.

4.4   HARDWARE-IN-THE-LOOP TEST RESULTS

In this section, to validate and compare the control performance of the proposed approaches, the HiL experiments were carried out with the new BBW system designed in Section 4.2. The dSPACE Autobox serves as a real-time platform, in which the models of vehicle motion, electric motor and battery are running. Meanwhile, the real braking system including hydraulic circuits, master and wheel cylinders, a vacuum booster and brake pedal, is equipped in the test bench. Several key parameters of the models used in tests are listed in Table 4.1.

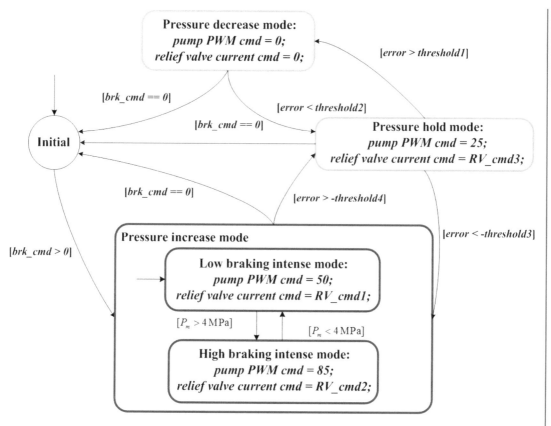

Figure 4.8: Schematic diagram of the switch coordinated control strategy for the relief valve and the pump-motor. *Error*: the difference between actual and target wheel cylinder pressures; *brk_cmd*: driver braking intention, which could be interpreted from the pressure in master cylinder; *Pm*: the pressure of master cylinder; *pump PWM cmd*: the *PWM* command value of pump-motor; *relief valve current cmd*: the current command value for relief valves.

The Freescale BA13 is used as the current control chip to generate a stably controlled coil current. The proposed hydraulic control algorithms are implemented in a real-time controller. The controller communicates with the dSPACE simulation platform through CAN (controller area network) bus. Figure 4.9 shows the overall experimental setup of the test bench.

First, the tracking performances of a ramp input under HPBPM and CLPDL control are compared. Next, a normal regenerative braking scenario is used to test the feasibility of the CLPDL control algorithm through actual implementation.

Table 4.1: Key parameters of the vehicle

Parameter	Value	Unit
Total mass (m)	1360	kg
Wheel base (L)	2.50	m
Coefficient of air resistance (C_D)	0.32	-
Nominal radius of tire (r)	0.295	m
Gear ratio	7.881	-

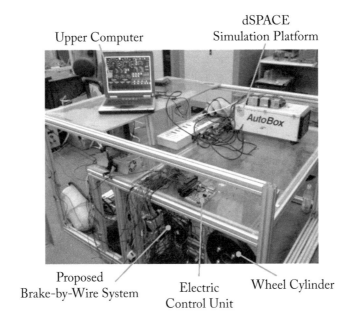

Figure 4.9: HiL test bench control algorithm tests.

4.4.1 COMPARISON OF HPBPM AND CLPDL CONTROL

So as to validate the performance of the designed modulation algorithms, the HiL testing scenario is set as: a normal deceleration is taken the operation condition; the objective value of the master cylinder pressure is set as a ramp input stabilizing at 3 MPa to mimic the normal braking intensity, and the adhesion coefficient is set at a high value of 0.8, indicating a dry surface.

The HiL test results with HPBPM are shown in Fig. 4.10a. At 0.3 s, the target master cylinder pressure starts to increase and the hydraulic pump driven by the electric motor builds up the pressure rapidly during the time period from 0.3–1.3 s. However, as the hydraulic pump requires 200–300 ms to establish negative pressure, there exists a short delay before pressure

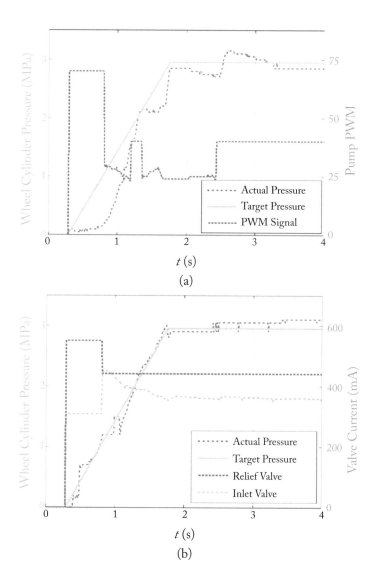

Figure 4.10: (a) HiL test results of the hydraulic pump based pressure modulation for the pressure tracking scenario; and (b) HiL test results of the CLPDL for the pressure tracking scenario.

Table 4.2: Comparison of pressure tracking performance under conventional PWM, HPBPM, and CLPDL control methods

	Conventional PWM	HPBPM	CLPDL
Average absolute value of the pressure tracking error (MPa)	0.1665	0.2284	0.0875
RMS (MPa)	0.2378	0.3291	0.1049

actually begins to build up at the beginning of braking. This delay degrades the pressure tracking performance. It was observed that the value of the PWM command decreases as the difference between actual and target pressures decrease. There are two steps in the PWM signal for the pump-motor: one is from 1.3–1.5 s, and the other is from 2.4 s to the end. The reason for the two steps is that the system is in the pressure hold mode and the pump is operating at a constant speed to maintain the pressure level of the whole braking circuit, thereby ensuring the responsiveness for the next pressure increase mode.

The same HiL test scenario is implemented to test the CLPDL control algorithm. The experimental data are presented in Fig. 4.10b. As the target wheel pressure increases, the coil current of the inlet valve jumps from 0–300 mA, allowing the brake fluid to flow into the wheel cylinder. The coil current value of the relief valve remains constant, thereby cooperating with the hydraulic pump to achieve a stable upstream pressure of the inlet valve. With the difference between the target and actual wheel cylinder pressures decreasing, the system transitions into pressure hold mode, and the value of the coil current is modulated to be smaller, thereby fully closing the inlet valve.

The superior tracking performance of the CLPDL control algorithm is evident from the comparison with the HPBPM control algorithm. In addition, the CLPDL control algorithm, in which there is no noise generated from the valves, provides more advantages over HPBPM control, in which PWM control for the pump-motor and inlet valve is used. The tracking performances of the target wheel cylinder pressure with HPBPM, CLPDL and conventional PWM control, which are adopted in [75, 92, 93], are compared quantitatively in Table 4.2. The average absolute value and RMS of pressure tracking error with CLPDL control are 0.0875 MPa and 0.1049 MPa, which are 61.69% and 68.13%, respectively, better than those with HPBPM and 47.45% and 55.89%, respectively, better than those with conventional PWM control. This finding indicates the enhancement of pressure modulation accuracy with the CLPDL control method.

According to the above-discussed HiL experimental results, with HPBPM, although the system noise is reduced compared to that with conventional PWM control for the inlet valve, the tracking performance of the target pressure is not guaranteed. The average absolute value and RMS of pressure tracking error with HPBPM control are 37.18% and 38.39%, respectively, which are worse than those with conventional PWM control. Thus, from both perspectives of control accuracy and noise reduction, the newly proposed CLPDL control appears to be an effective method to address the wheel cylinder pressure modulation problem.

4.4.2   BRAKE BLENDING ALGORITHM BASED ON CLPDL MODULATION

The brake blending performance between regenerative braking and hydraulic braking has a significant effect upon the regeneration efficiency and brake safety. Thus, the requirement of hydraulic pressure modulation accuracy is crucial. In addition, unlike emergency conditions, such as anti-lock braking, brake comfort and noise level, are also evaluated during regenerative braking, making the hydraulic pressure modulation even more difficult. To verify the performance of the CLPDL algorithm in real-world applications, a typical brake blending test of EVs is conducted on the HiL test bench. The set-up of the testing scenarios is the same as the one adopted in the tracking performance test. The brake maneuver is initialized with the vehicle speed of 72 km/h. The target vehicle is a front-wheel driven one, which equips with a centralized electric motor. The same brake blending control strategy as introduced in [94, 95, 98] is employed. The only modification is the replacement of the conventional PWM valve control with the CLPDL control algorithm. The details of the brake blending algorithm are presented in [94, 95, 98].

The HiL test results for the front axle are shown in Fig. 4.11a. At approximately 0.8 s, the target master cylinder pressure begins to increase. The electric motor applies the braking torque gradually on the front axle. According to the brake blending strategy, the hydraulic braking, which is under the control of the CLPDL algorithm, decreases when the regenerative braking torque increases. After 1.5 s, the target value of master cylinder pressure remains stable at approximately 3 MPa, and accordingly the hydraulic brake pressure is modulated dynamically under CLPDL control.

During the pressure increase phase, the outlet valve of the brake cylinder is not energized. Starting from 6.5 s, the regenerative braking torque gradually decreases and finally down to zero because of the characteristics of the electric motor. Thus, the hydraulic brake pressure needs to be increased accordingly to fulfill the brake demand. Under the control of CLPDL, the hydraulic braking pressure changes smoothly and the driver's braking intension is well satisfied.

The HiL tests results for brake cylinders in rear axle are shown in Fig. 4.11b. Since there is no regenerative braking applied on the rear axle, the control task is same as those illustrated in Fig. 4.11a. As shown in Fig. 4.11b, the target pressure of the rear wheel cylinder is well-tracked. The overall experimental results show that the target brake pressure is followed very well through valve modulation with the proposed approach, and the blended braking brakes at the front wheels are coordinated well, thereby validating the effectiveness of the CLPDL control algorithm of the hydraulic brake pressure.

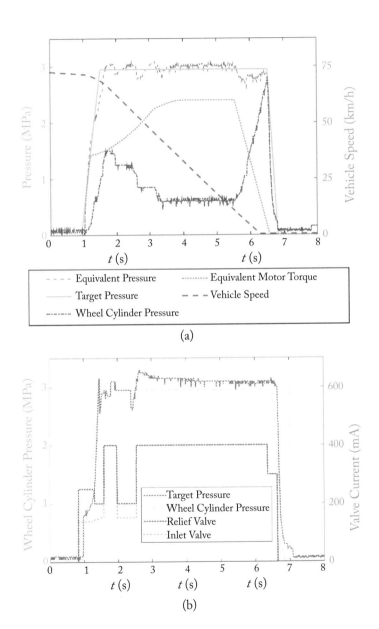

Figure 4.11: (a) HiL test results of the CLPDL-based brake blending control strategy for the front axle; and (b) HiL test results of the CLPDL-based brake blending control strategy for the rear axle.

CHAPTER 5

Conclusions

This book introduces the design optimization, state estimation, and advanced control methods for CPVS and their applications in real-world automotive systems.

In Chapter 2, a CPS-based framework for co-design optimization of an automated electric vehicle with different driving styles was proposed. The multi-objective optimization problem was formulated. The driving style recognition algorithm was developed using unsupervised machine learning and validated via vehicle testing. The system modeling and experimental verification were carried out. Vehicle control algorithms were synthesized for three typical driving styles with different protocol selections. The performance exploration methodology and algorithms were proposed. Test results show that the overall performances of the vehicle were significantly improved by the proposed co-design optimization approach. Future work will be focused on real vehicle application of the proposed methods and CPS design methodology improvement.

In Chapter 3, a novel probabilistic estimation method of brake pressure is developed for a safety critical CPVS based on multilayer ANN with LMBP training algorithm. The high-level architecture of the proposed multilayer ANN for brake pressure estimation is illustrated at first. Then, the standard BP algorithm used for training of FFNN is introduced. Based on the basic concept of BP, a more efficient algorithm of LMBP method is developed for model training. The real vehicle testing is carried out on a chassis dynamometer under NEDC driving cycles. Experimental data of the vehicle and powertrain systems is collected, and feature vectors for FFNN training collection are selected. With the vehicle data obtained, the developed multilayer ANN is trained. The experimental results show that the developed ANN model, which is trained by LMBP, can accurately estimate the brake pressure, and its performance is advantageous over other learning-based methods with respect to estimation accuracy, demonstrating the feasibility and effectiveness of the proposed algorithm. Further work can be carried out in the following areas: the proposed algorithm will be further refined with onboard road testing; intelligent control algorithms of braking system will be designed based on state estimation.

High-precision control of mechatronics is an important foundation of the development of safe, smart and sustainable CPVS [99–104]. In Chapter 4, a typical safety-critical CPVS, i.e., the BBW system, was introduced. Compared to the existing BBW system, the newly developed system enjoys the advantage of a simple structure and low cost because only conventional valves and sensors are added to the usual hydraulic layouts. Two pressure modulation methods, namely, the HPBPM and CLPDL modulation, were proposed to improve the modulation precision of hydraulic brake pressure and reduce valve's operation noise as well. Experiments were conducted in HiL test rig to demonstrate the performance of the proposed control methods. Experimental results showed that, in spite of the reduction in noise in the HPBPM control method, when compared to the conventional PWM control, control accuracy is not guaranteed. The CLPDL method achieves both good performance of tracking target pressure and reduced noise. To further validate the feasibility of the newly proposed BBW system and the CLPDL control method, a typical regenerative braking scenario was utilized. The CLPDL control method was implemented within a regenerative braking strategy. The HiL test results illustrated that both the front and rear wheel cylinders fulfilled the driver braking intention and that the regenerative braking and hydraulic braking cooperate well with each other. For the future work, real vehicle tests will be conducted, and qualitative comparisons between proposed BBW system and existed BBW systems will be explored.

References

[1] Petrovskaya, A. and Thrun, S. Model based vehicle detection and tracking for autonomous urban driving, *Autonomous Robots 26*, 2–3:123–139, 2009. DOI: 10.1007/s10514-009-9115-1. 1

[2] González, D., Pérez, J., et al. A review of motion planning techniques for automated vehicles, *IEEE Transactions on Intelligent Transportation Systems*, 17(4):1135–1145, 2016. DOI: 10.1109/tits.2015.2498841. 1

[3] Lv, C., Hu, X., Sangiovanni-Vincentelli, A., Li, Y., Martinez, C. M., and Cao, D. Driving-style-based codesign optimization of an automated electric vehicle: A cyber-physical system approach, *IEEE Transactions on Industrial Electronics*, 66(4):2965–2975, 2018. DOI: 10.1109/tie.2018.2850031.

[4] Lv, C., Cao, D., et al. Analysis of autopilot disengagements occurring during autonomous vehicle testing, *IEEE/CAA Journal of Automatica Sinica*, 5:58–68, 2018. DOI: 10.1109/jas.2017.7510745.

[5] Shen, C., Shi, Y., and Buckham, B. Integrated path planning and tracking control of an AUV: A unified receding horizon optimization approach, *IEEE/ASME Transactions on Mechatronics 22*, 3:1163–1173, 2017. DOI: 10.1109/tmech.2016.2612689. 1

[6] Lv, C., Wang, H., Cao, D., Zhao, Y., Auger, D. J., Sullman, M., et al. Characterization of driver neuromuscular dynamics for human-automation collaboration design of automated vehicles, *IEEE/ASME Transactions on Mechatronics*, 23(6), pp. 2558–2567, 2018. DOI: 10.1109/tmech.2018.2812643. 1

[7] Ji, X., Liu, Y., et al. Interactive control paradigm-based robust lateral stability controller design for autonomous automobile path tracking with uncertain disturbance: A dynamic game approach, *IEEE Transactions on Vehicular Technology*, 67(8), pp. 6906–6920, 2018. DOI: 10.1109/tvt.2018.2834381.

[8] Li, H., Shi, Y., and Yan, W. Distributed receding horizon control of constrained nonlinear vehicle formations with guaranteed γ-gain stability, *Automatica*, 68:148–154, 2016. DOI: 10.1016/j.automatica.2016.01.057.

[9] Zhao, W., Qin, X., and Wang, C. Yaw and lateral stability control of automotive four-wheel steer-by-wire system, *IEEE/ASME Transactions on Mechatronics*, 23(6), pp. 2628–2637, 2018. DOI: 10.1109/tmech.2018.2812220. 1

[10] Katzourakis, D. I., Abbink, D. A., et al. Steering force feedback for human-machine-interface automotive experiments, *IEEE Transactions on Instrumentation and Measurement*, 60:32–43, 2011. DOI: 10.1109/tim.2010.2065550. 1

[11] Miedl, F. and Tille, T. 3D surface-integrated touch-sensor system for automotive HMI applications, *IEEE/ASME Transactions on Mechatronics*, 21:787–794, 2016. DOI: 10.1109/tmech.2015.2466455. 1

[12] Kapania, N. R. and Gerdes, J. C. Design of a feedback-feedforward steering controller for accurate path tracking and stability at the limits of handling, *Vehicle System Dynamics*, 53:1687–1704, 2015. DOI: 10.1080/00423114.2015.1055279. 1

[13] Wang, R. and Lukic, S. Review of driving conditions prediction and driving style recognition based control algorithms for hybrid electric vehicles, *Vehicle Power and Propulsion Converence*, pages 1–7, 2011. DOI: 10.1109/vppc.2011.6043061. 1

[14] Kesting, A., Treiber, M., and Helbing, D. Enhanced intelligent driver model to access the impact of driving strategies on traffic capacity, *Philosophical Transactions of the Royal Society of London A: Mathematical, Physical and Engineering Sciences*, 368:4585–4605, 2010. DOI: 10.1098/rsta.2010.0084. 1

[15] Kesting, A., Treiber, M., Schönhof, M., and Helbing, D. Extending adaptive cruise control to adaptive driving strategies, *Transportation Research Record: Journal of the Transportation Research Board*, 2000:16–24, 2007. DOI: 10.3141/2000-03. 1

[16] Lv, C., Xing, Y., et al. Hybrid-learning-based classification and quantitative inference of driver braking intensity of an electrified vehicle, *IEEE Transactions on Vehicular Technology*, 67(7), pp. 5718–5729, 2018. DOI: 10.1109/tvt.2018.2808359. 1

[17] Zhang, J., Lv, C., Gou, J., and Kong, D. Cooperative control of regenerative braking and hydraulic braking of an electrified passenger car, *Proc. of the Institution of Mechanical Engineers, Part D: Journal of Automobile Engineering*, 226:1289–1302, 2012. DOI: 10.1177/0954407012441884. 7, 10, 11, 12, 13, 44

[18] Zhao, W., Zhang, H., and Li, Y. Displacement and force coupling control design for automotive active front steering system, *Mechanical Systems and Signal Processing*, 106:76–93, 2018. DOI: 10.1016/j.ymssp.2017.12.037. 1, 10

[19] Neubauer, J. and Wood, E. Accounting for the variation of driver aggression in the simulation of conventional and advanced vehicles, *SAE Technical Paper*, 01:1453, 2013. DOI: 10.4271/2013-01-1453. 1

[20] Li, L., You, S., Yang, C., Yan, B., Song, J., and Chen, Z. Driving-behavior-aware stochastic model predictive control for plug-in hybrid electric buses, *Applied Energy*, 162:868–897, 2016. DOI: 10.1016/j.apenergy.2015.10.152. 1

[21] Derler, P., Lee, E. A., et al. Modeling cyber-physical systems, *Proc. of the IEEE*, 100(1):13–28, January 2012. DOI: 10.1109/jproc.2011.2160929. 5

[22] Sangiovanni-Vincentelli, A., Damm, W., and Passerone, R. Taming Dr. Frankenstein: Contract-based design for cyber-physical systems, *European Journal of Control*, 18:217–238, 2012. DOI: 10.3166/ejc.18.217-238.

[23] Lv, C., Xing, Y., et al. Levenberg-Marquardt backpropagation training of multi-layer neural networks for state estimation of a safety critical cyber-physical system, *IEEE Transactions on Industrial Informatics*, 14(8), pp. 3436–3446, 2017. DOI: 10.1109/tii.2017.2777460.

[24] Nuzzo, P., Sangiovanni-Vincentelli, A., et al. A platform-based design methodology with contracts and related tools for the design of cyber-physical systems, *Proc. IEEE*, 103(11):2104–2132, 2015. DOI: 10.1109/jproc.2015.2453253. 2

[25] Lv, C., Liu, Y., et al. Simultaneous observation of hybrid states for cyber-physical systems: A case study of electric vehicle powertrain, *IEEE Transactions on Cybernetics*, 48(8), pp. 2357–2367, 2018. DOI: 10.1109/TCYB.2017.2738003. 2, 11, 13

[26] Bradley, J. M. and Atkins, E. M. Cyber-physical optimization for unmanned aircraft systems, *Journal of Aerospace Information Systems*, 11(1):48–60, 2014. DOI: 10.2514/1.i010105. 2

[27] Hu, X., Wang, H., and Tang, X. Cyber-physical control for energy-saving vehicle following with connectivity, *IEEE Transactions on Industrial Electronics*, 2017. DOI: 10.1109/tie.2017.2703673. 2

[28] Li, Y., Lv, C., et al. High-precision modulation of a safety-critical cyber-physical system: Control synthesis and experimental validation, *IEEE/ASME Transactions on Mechatronics*, 23(6), pp. 2599–2608, 2018. DOI: 10.1109/tmech.2018.2833542. 2, 7

[29] Lv, C., Wang, H., et al. Cyber-physical system based optimization framework for intelligent powertrain control, *SAE International Journal of Commercial Vehicles*, 10:254–264, 2017. DOI: 10.4271/2017-01-0426. 3, 7, 8, 10

[30] Finn, J., Nuzzo, P., and Sangiovanni-Vincentelli, A. A mixed discrete-continuous optimization scheme for cyber-physical system architecture exploration. *Proc. of the IEEE/ACM International Conference on Computer-Aided Design*, pages 216–223, 2015. DOI: 10.1109/iccad.2015.7372573.

[31] Lv, C., Zhang, J., Nuzzo, P., Sangiovanni-Vincentelli, A., et al. Design optimization of the control system for the powertrain of an electric vehicle: A cyber-physical system approach, *Mechatronics and Automation, IEEE International Conference on*, pages 814–819, 2015. DOI: 10.1109/icma.2015.7237590. 3

[32] Gilman, E., Keskinarkaus, A., Tamminen, S., et al. Personalised assistance for fuel-efficient driving, *Transportation Research Part C: Emerging Technologies*, 58:681–705, 2015. DOI: 10.1016/j.trc.2015.02.007. 7

[33] Martinez, C. M., Heucke, M., Wang, F. Y., et al. Driving style recognition for intelligent vehicle control and advanced driver assistance: A survey, *IEEE Transactions on Intelligent Transportation Systems*, 2017. DOI: 10.1109/tits.2017.2706978. 7

[34] Martínez, C. M. iHorizon-enabled energy management for plug-in hybrid electric vehicles, Ph.D. dissertation, Dept. Auto. Eng., Cranfield University, Bedford, UK, 2017. 7, 8, 9

[35] Mitschke, M. and Wallentowitz, H. *Dynamik der Kraftfahrzeuge*, 4th ed., Springer Verlag, Berlin, 2004. DOI: 10.1007/978-3-662-11585-5. 11, 15

[36] Gao, Y. and Ehsani, M. Electronic braking system of EV and HEV—Integration of regenerative braking, automatic braking force control and ABS, *SAE Technical Paper*, 01:2478, 2001. DOI: 10.4271/2001-01-2478. 10, 11

[37] Pacejka, H. B. and Bakker, E. The magic formula tyre model, *Vehicle System Dynamics*, 21(S):1–18, 1992. DOI: 10.1080/00423119208969994. 12

[38] Song, B. and Hedrick, J. K. *Dynamic Surface Control of Uncertain Nonlinear Systems: An LMI Approach*, Springer, 2011. DOI: 10.1007/978-0-85729-632-0. 14

[39] Fazeli, A., Zeinali, M., and Khajepour, A. Application of adaptive sliding mode control for regenerative braking torque control, *IEEE/ASME Transactions on Mechatronics*, 17:745–755, 2012. DOI: 10.1109/tmech.2011.2129525. 15

[40] Alam, B., Wu, J., Wang, G., and Cao, J. Sensing and decision making in cyber-physical systems: The case of structural event monitoring, *IEEE Transactions on Industrial Informatics*, 2016. DOI: 10.1109/tii.2016.2518642. 23

[41] Wang, F.-Y. The emergence of intelligent enterprises: From CPS to CPSS, *IEEE Intelligent Systems*, 25(4):85–88, 2010. DOI: 10.1109/mis.2010.104.

[42] Gong, H., Li, R., An, J., Chen, W., and Li, K. Scheduling algorithms of flat semidormant multi-controllers for a cyber-physical system, *IEEE Transactions on Industrial Informatics*, 13(4), pp. 1665–1680, 2017. DOI: 10.1109/tii.2017.2690939.

[43] Wang, F.-Y. Control 5.0: From Newton to Merton in popper's cyber-social-physical spaces, *IEEE/CAA Journal of Automatica Sinica*, 3(3):233–234, 2016. DOI: 10.1109/jas.2016.7508796. 23

[44] Zhou, Q., Zhang, W., Cash, S., Olatunbosun, O., Xu, H., and Lu, G. Intelligent sizing of a series hybrid electric power-train system based on chaos-enhanced accelerated particle swarm optimization, *Applied Energy*, 189:588–601, 2017. DOI: 10.1016/j.apenergy.2016.12.074. 23

[45] Lv, C., Wang, H., and Cao, D. High-precision hydraulic pressure control based on linear pressure-drop modulation in valve critical equilibrium state, *IEEE Transactions on Industrial Electronics*, 2017. DOI: 10.1109/tie.2017.2694414. 49, 51, 52, 53

[46] Kisacikoglu, M., Erden, F., and Erdogan, N. Distributed control of PEV charging based on energy demand forecast, *IEEE Transactions on Industrial Informatics*, 14(1), pp. 332–341, 2017. DOI: 10.1109/tii.2017.2705075.

[47] Qin, Y., Langari, R., Wang, Z., Xiang, C., and Dong, M. Road excitation classification for semi-active suspension system with deep neural networks, *Journal of Intelligent and Fuzzy Systems Preprint*, pages 1–12, 2017. DOI: 10.3233/jifs-161860.

[48] Wang, S., Dong, Z. Y., et al. Stochastic collaborative planning of electric vehicle charging stations and power distribution system, *IEEE Transactions on Industrial Informatics*, 14(1), pp. 321–331, 2017. DOI: 10.1109/pesgm.2016.7741442.

[49] Xing, Y. and Lv, C. Dynamic state estimation for the advanced brake system of electric vehicles by using deep recurrent neural networks, *IEEE Transactions on Industrial Electronics*, 2019. DOI: 10.1109/tie.2019.2952807.

[50] Mirzaei, M. J., Kazemi, A., and Homaee, O. A probabilistic approach to determine optimal capacity and location of electric vehicles parking lots in distribution networks, *IEEE Transactions on Industrial Informatics*, 12(5):1963–1972, 2016. DOI: 10.1109/tii.2015.2482919. 23

[51] Lv, C., Zhang, J., and Li, Y. Extended-Kalman-filter-based regenerative and friction blended braking control for electric vehicle equipped with axle motor considering damping and elastic properties of electric powertrain, *Vehicle System Dynamics*, 52(11):1372–1388, 2014. DOI: 10.1080/00423114.2014.938663. 23

[52] Zhao, B., Lv, C., and Hofman, T. *Automotive Innovations*, 2(2):146–156, 2019. https://doi.org/10.1007/s42154--019-00059-z 23

[53] Xing, Y., Lv, C., et al. Driver lane change intention inference for intelligent vehicles: Framework, survey, and challenges, *IEEE Transactions on Vehicular Technology*, 68(5):4377–4390, 2019. DOI: 10.1109/tvt.2019.2903299. 23

[54] Hu, X., Moura, S. J., Murgovski, N., Egardt, B., and Cao, D. Integrated optimization of battery sizing, charging, and power management in plug-in hybrid electric vehicles, *IEEE Transactions on Control Systems Technology*, 24(3):1036–1043, 2016. DOI: 10.1109/tcst.2015.2476799.

[55] Cena, G., Bertolotti, I. C., et al. CAN with eXtensible in-frame reply: Protocol definition and prototype implementation, *IEEE Transactions on Industrial Informatics*, 2017 (in press). DOI: 10.1109/tii.2017.2714183.

[56] Lv, C., Zhang, J., Li, Y., and Yuan, Y. Novel control algorithm of braking energy regeneration system for an electric vehicle during safety—critical driving maneuvers, *Energy Conversion and Management*, 106:520–529, 2015. DOI: 10.1016/j.enconman.2015.09.062. 23

[57] Martins, L. and Gorschek, T. Requirements engineering for safety-critical systems: Overview and challenges, *IEEE Software*, 2017. DOI: 10.1109/ms.2017.265100352. 23

[58] Ames, A. D., Xu, X., Grizzle, J. W., and Tabuada, P. Control barrier function based quadratic programs for safety critical systems, *IEEE Transactions on Automatic Control*, 2016. DOI: 10.1109/tac.2016.2638961. 23

[59] Shoukry, Y., Nuzzo, P., Puggelli, A., Sangiovanni-Vincentelli, A. L., et al. Secure state estimation for cyber physical systems under sensor attacks: A satisfiability modulo theory approach, *IEEE Transactions on Automatic Control*, 2017. DOI: 10.1109/tac.2017.2676679. 23

[60] Ding, N. and Zhan, X. Model-based recursive least square algorithm for estimation of brake pressure and road friction, *Proc. of the FISITA 2012 World Automotive Congress*, pages 137–145, Springer, Berlin, Heidelberg, 2013. DOI: 10.1007/978-3-642-33795-6_12. 24

[61] Jiang, G., Miao, X., Wang, Y., et al. Real-time estimation of the pressure in the wheel cylinder with a hydraulic control unit in the vehicle braking control system based on the extended Kalman filter, *Proc. of the Institution of Mechanical Engineers, Part D: Journal of Automobile Engineering*, 2016 (in press). DOI: 10.1177/0954407016671685. 24

[62] Li, L., Song, J., and Han, Z. Hydraulic model and inverse model for electronic stability program online control system, *Chinese Journal of Mechanical Engineering*, 44(2):139, 2008. DOI: 10.3901/jme.2008.02.139. 24

[63] Zhang, J., Lv, C., Yue, X., Qiu, M., Gou, J., and He, C. Development of the electrically-controlled regenerative braking system for electrified passenger vehicle, *SAE Technical Paper*, 01:1463, 2013. DOI: 10.4271/2013-01-1463. 24

[64] Yao, J., Zhang, Y., and Wang, J. Research on algorithm of braking pressure estimating for anti-lock braking system of motorcycle, *Aircraft Utility Systems (AUS), IEEE International Conference on*, pages 586–591, 2016. DOI: 10.1109/aus.2016.7748118. 24

[65] O'Dea, K. Anti-lock braking performance and hydraulic brake pressure estimation, *SAE Technical Paper*, 01:1061, 2005. DOI: 10.4271/2005-01-1061. 24

[66] Demuth, H. B., Beale, M. H., De Jess, O., and Hagan, M. T. Neural network design: Martin Hagan, Sep. 2014. 26

[67] Soualhi, A., Makdessi, M., German, R., et al. Heath monitoring of capacitors and super-capacitors using neo fuzzy neural approach, *IEEE Transactions on Industrial Informatics*, 2017. DOI: 10.1109/tii.2017.2701823. 27

[68] Bishop, C. M. *Pattern Recognition and Machine Learning*, Springer, 2006. 30

[69] Dreyfus, G. *Neural Networks: Methodology and Applications*, Springer Science and Business Media, 2005. 30

[70] Hagan, M. T. and Menhaj, M. B. Training feedforward networks with the Marquardt algorithm, *IEEE Transactions on Neural Networks*, 5(6):989–993, 1994. DOI: 10.1109/72.329697. 31

[71] Lv, C., Zhang, J., Li, Y., and Yuan, Y. Mechanism analysis and evaluation methodology of regenerative braking contribution to energy efficiency improvement of electrified vehicles, *Energy Conversion and Management*, 92:469–482, 2015. DOI: 10.1016/j.enconman.2014.12.092. 34

[72] Refaeilzadeh, P., Tang, L., and Liu, H. Cross-validation, *Encyclopedia of Database Systems*, pages 532–538, Springer 2009. DOI: 10.1007/978-1-4899-7993-3_565-2. 36

[73] Budescu, D. V. Dominance analysis: A new approach to the problem of relative importance of predictors in multiple regression, *Psychological Bulletin*, 114(3):542, 1993. DOI: 10.1037/0033-2909.114.3.542. 40

[74] Zhou, Q., Zhang, Y., et al. Cyber-physical energy-saving control for hybrid aircraft-towing tractor based on online swarm intelligent programming, *IEEE Transactions on Industrial Informatics*, 2017. DOI: 10.1109/tii.2017.2781230. 43

[75] Zhang, J., Li, Y., Lv, C., and Yuan, Y. New regenerative braking control strategy for rear-driven electrified minivans, *Energy Conversion and Management*, 82:135–145, 2014. DOI: 10.1016/j.enconman.2014.03.015. 44, 58

[76] Hu, J. S., Wang, Y., Fujimoto, H., and Hori, Y. Robust yaw stability control for in-wheel motor electric vehicles, *IEEE/ASME Transactions on Mechatronics*, 2017. DOI: 10.1109/tmech.2017.2677998. 43

[77] Li, W., Lv, C., Gopalswamy, S., Li, L., and Khajepour, A. Guest editorial focused section on mechatronics in cyber-physical systems, *IEEE/ASME Transactions on Mechatronics*, 23(6):2533–2536, 2018. DOI: 10.1109/tmech.2018.2853596. 43

[78] Yu, L. and Chang, T. Zero vibration on-off position control of dual solenoid actuator, *IEEE Transactions on Industrial Electronics*, 57(7):2519–2526, 2010. DOI: 10.1109/iecon.2007.4460165.

[79] Munoz, F., Li, W., et al. Analysis of magnetic interaction in remotely controlled magnetic devices and its application to a capsule robot for drug delivery, *IEEE/ASME Transactions on Mechatronics*, 2017. DOI: 10.1109/tmech.2017.2764902.

[80] Tang, X., Du, H., Sun, S., Ning, D., Xing, Z., and Li, W. Takagi–Sugeno fuzzy control for semi-active vehicle suspension with a magnetorheological damper and experimental validation, *IEEE/ASME Transactions on Mechatronics*, 22(1):291–300, 2017. DOI: 10.1109/tmech.2016.2619361. 43

[81] Xing, Y., Lv, C., Huaji, W., Wang, H., and Cao, D. Recognizing driver braking intention with vehicle data using unsupervised learning methods, *SAE Technical Paper*, 01:0433, 2017. DOI: 10.4271/2017-01-0433. 43

[82] Oshima, T., Fujiki, N., Nakao, S., and Kimura, T. Development of an electrically driven intelligent brake system, *SAE International Journal of Passenger Cars*, 4(1):399–405, 2011. DOI: 10.4271/2011-01-0568.

[83] Milanes, V., Gonzalez, C., Naranjo, E., Onieva, E., and De Pedro, T. Electrohydraulic braking system for autonomous vehicles, *International Journal of Automotive Technology*, 11(1):89–95, 2010. DOI: 10.1007/s12239-010-0012-6. 43

[84] Rossa, C., Jaegy, A., Lozada, J., and Micaelli, A. Design considerations for magnetorheological brakes, *IEEE/ASME Transactions on Mechatronics*, 19(5):1669–1680, 2014. DOI: 10.1109/tmech.2013.2291966. 43

[85] Bae, J., Kim, Y., Son, Y., Moon, H., Yoo, C., and Lee, J. Self-excited induction generator as an auxiliary brake for heavy vehicles and its analog controller, *IEEE Transactions on Industrial Electronics*, 62(5):3091–3100, 2015. DOI: 10.1109/tie.2014.2379218.

[86] Saric, S., Bab-Hadiasher, A., and Hoseinnezhad, R. Clamp-force estimation for a brake-by-wire system: A sensor-fusion approach, *IEEE Transactions on Vehicle Technology*, 57(2):778–786, 2008. DOI: 10.1109/tvt.2007.905251. 43

[87] Champagne, R. and Stephens, S. Optimizing valve actuator parameters to enhance control valve performance, *ISA Transactions*, 35(3):217–23, 2005. DOI: 10.1016/s0019-0578(96)00028-6. 43

[88] Choi, S. and Cho, D. Control of wheel slip ratio using sliding mode controller with pulse width modulation, *Vehicle System Dynamics*, 32:267–284, 1999. DOI: 10.1076/vesd.32.4.267.2080. 44

[89] Ahn, K. and Yokota, S. Intelligent switching control of pneumatic actuator using on/off solenoid valves, *Mechatronics*, 15(6):683–702, 2005. DOI: 10.1016/j.mechatronics.2005.01.001. 44

[90] Wang, W., Song, J., Li, L., and Li, H. High speed on-off solenoid valve with proportional control based on high frequency PWM control, *Journal of Tsinghua University (Science and Technology)*, 51(5), 2011. 44, 48

[91] Zhao, X., Li, L., Song, J., Li, C., and Gao, X. Linear control of switching valve in vehicle hydraulic control unit based on sensorless solenoid position estimation, *IEEE Transactions on Industrial Electronics*, 63(7), pp. 4073–4085, 2017. DOI: 10.1109/tie.2016.2541080. 44

[92] Xing, Y., Lv, C., Wang, H., Cao, D., Velenis, E., and Wang, F. Y. Driver activity recognition for intelligent vehicles: A deep learning approach, *IEEE Transactions on Vehicular Technology*, 2019. DOI: 10.1109/tvt.2019.2908425. 44, 58

[93] Lv, C., et al. Brake-blending control of EVs, *Modeling, Dynamics and Control of Electrified Vehicles*, pages 275–308, 2018. DOI: 10.1016/b978-0-12-812786-5.00008-2. 58

[94] Zhang, J., Lv, C., Yue, X., Li, Y., and Yuan, Y. Study on a linear relationship between limited pressure difference and coil current of on/off valve and its influential factors, *ISA Transactions*, 53(1):150–161, 2014. DOI: 10.1016/j.isatra.2013.09.008. 59

[95] Lv, C., Zhang, J., Li, Y., Sun, D., and Yuan, Y. Hardware-in-the-loop simulation of pressure-difference-limiting modulation of the hydraulic brake for regenerative braking control of electric vehicles, *Proc. of the Institution of Mechanical Engineers, Part D: Journal of Automobile Engineering*, 228(6):649–662, 2014. DOI: 10.1177/0954407013516942. 44, 48, 59

[96] Soga, M., Shimada, M., Sakamoto, J., and Otomo, A. Development of vehicle dynamics management system for hybrid vehicles—ECB system for improved environmental and vehicle dynamic performance, *SAE Technical Paper*, 01:1586, 2002. DOI: 10.4271/2002-01-1586. 44, 45

[97] van Zanten, A. Bosch ESP systems: 5 years of experience, *SAE Technical Paper*, 01:1633, 2000. DOI: 10.4271/2000-01-1633. 45

[98] Zhang, J., Li, Y., Lv, C., Gou, J., and Yuan, Y. Time-varying delays compensation algorithm for powertrain active damping of an electrified vehicle equipped with an axle

motor during regenerative braking, *Mechanical Systems and Signal Processing*, 2017. DOI: 10.1016/j.ymssp.2015.10.008. 59

[99] Xing, Y., Lv, C., Wang, H., Cao, D., and Velenis, E. Dynamic integration and online evaluation of vision-based lane detection algorithms, *IET Intelligent Transport Systems*, 13(1):55–62, 2018. DOI: 10.1049/iet-its.2018.5256. 61

[100] Lv, C., Wang, H., Cao, D., Zhao, Y., Auger, D. J., Sullman, M., Matthias, R., Skrypchuk, L., and Mouzakitis, A. Characterization of driver neuromuscular dynamics for human—automation collaboration design of automated vehicles, *IEEE/ASME Transactions on Mechatronics*, 23(6):2558–2567, 2018. DOI: 10.1109/tmech.2018.2812643.

[101] Xing, Y., Lv, C., Chen, L., Wang, H., Wang, H., Cao, D., Velenis, E., and Wang, F. Y., Advances in vision-based lane detection: Algorithms, integration, assessment, and perspectives on ACP-based parallel vision, *IEEE/CAA Journal of Automatica Sinica*, 5(3):645–661, 2018. DOI: 10.1109/jas.2018.7511063.

[102] Xing, Y., Lv, C., Zhang, Z., Wang, H., Na, X., Cao, D., Velenis, E., and Wang, F. Y. Identification and analysis of driver postures for in-vehicle driving activities and secondary tasks recognition, *IEEE Transactions on Computational Social Systems*, 5(1):95–108, 2017. DOI: 10.1109/tcss.2017.2766884.

[103] Xing, Y., Lv, C., Cao, D., Wang, H., and Zhao, Y. Driver workload estimation using a novel hybrid method of error reduction ratio causality and support vector machine, *Measurement*, 114:390–397, 2018. DOI: 10.1016/j.measurement.2017.10.002.

[104] Lv, C., Zhang, J., Li, Y., and Yuan, Y. Directional-stability-aware brake blending control synthesis for over-actuated electric vehicles during straight-line deceleration, *Mechatronics*, 38:121–131, 2016. DOI: 10.1016/j.mechatronics.2015.12.010. 61

Authors' Biographies

CHEN LV

Chen Lv is currently an Assistant Professor of School of Mechanical and Aerospace Engineering, Nanyang Technological University. He is also a Cluster Director in Future Mobility Solutions at Energy Research Institute at NTU. He received his Ph.D. from Department of Automotive Engineering, Tsinghua University, China in 2016. He was a joint Ph.D. researcher at EECS Dept., University of California, Berkeley, USA during 2014-2015, and worked as a Research Fellow at Advanced Vehicle Engineering Center, Cranfield University, UK during 2016-2018. He joined NTU and founded the Automated Driving and Human-Machine System Research Group since June 2018. His research focuses on automated driving, human-machine intelligence, and intelligent electric vehicles, where he has contributed 2 book chapters, over 90 papers, and obtained 12 granted patents.

Dr. Lv serves an Academic Editor for *PLOS ONE*, *Automotive Innovation*, *International Journal of Electric and Hybrid Vehicles*, and *International Journal of Vehicle Systems Modelling and Testing*, and he was a Guest Editor for *IEEE/ASME Transactions on Mechatronics*, *IEEE Transactions on Industrial Informatics*, *Applied Energy*, and *International Journal of Powertrains*. He received the Highly Commended Paper Award of IMechE UK in 2012, the NSK Outstanding Mechanical Engineering Paper Award in 2014, the China SAE Outstanding Paper Award in 2015, the 1st Class Award of China Automotive Industry Scientific and Technological Invention in 2015, the Tsinghua University Graduate Student Academic Rising Star Nomination Award in 2015, the Tsinghua University Outstanding Doctoral Thesis Award in 2016, the Seal of Excellence of EU H2020 Marie Skłodowska-Curie Actions in 2017, the Best Workshop/Special Session Paper Award of IEEE Intelligent Vehicle Symposium in 2018, and CSAE Outstanding Doctoral Dissertation Award in 2018.

YANG XING

Yang Xing received his Ph.D. from Cranfield University, UK, in 2018. He is currently a research fellow with the department of mechanical and aerospace engineering at Nanyang Technological University, Singapore. His research interests include machine learning, driver behavior modeling, intelligent multi-agent collaboration, and intelligent/autonomous vehicles. His work focuses on the understanding of driver behaviors using machine-learning methods and intelligent and automated vehicle design. He received the IV2018 Best Workshop/Special Issue Paper Award. Dr. Xing serves as a Guest Editor for *IEEE Internet of Things*, and he is an active reviewer for *IEEE Transactions on Vehicular Technology*, *Industrial Electronics*, and *Intelligent Transportation Systems*.

JUNZHI ZHANG

Junzhi Zhang received his B.E. in Transportation Engineering and his M.S. and Ph.D. in Vehicle Engineering from the Jiliin University of Technology, Changchun, China in 1992, 1995, and 1997, respectively. From 1998–1999, Dr. Zhang was a Research Associate in Department of Automotive Engineering in Tsinghua University, Beijing, China. In 1999, Dr. Zhang joined Tsinghua University and founded the Hybrid Powertrain Systems Laboratory whose major research interests include modeling, control and diagnosis of hybrid, electric vehicle. Dr. Zhang become a full professor in Department of Automotive Engineering in Tsinghua University in 2008. Dr. Zhang is the author or co-author of more than 50 peer-reviewed publications and 20 Chinese patents.

DONGPU CAO

Dongpu Cao received his Ph.D. from Concordia University, Canada, in 2008. He is currently an Associate Professor at Mechanical and Mechatronics Engineering, University of Waterloo, Canada. His research focuses on vehicle dynamics and control, automated driving and parallel driving, where he has contributed more than 100 publications and 1 US patent. He received the ASME AVTT'2010 Best Paper Award and 2012 SAE Arch T. Colwell Merit Award. Dr. Cao serves as an Associate Editor for *IEEE Transactions on Intelligent Transportation Systems*, *IEEE Transactions on Vehicular Technology*, *IEEE Transactions on Industrial Electronics*, *IEEE/ASME Transactions on Mechatronics*, and *ASME Journal of Dynamic Systems, Measurement, and Control*. He has been a Guest Editor for *Vehicle System Dynamics*, and *IEEE Transactions on Human-Machine Systems*. He serves on the SAE International Vehicle Dynamics Standards Committee and a few ASME, SAE, and IEEE technical committees.

Printed in the United States
by Baker & Taylor Publisher Services